Lecture Notes in Artificial Intel

Edited by R. Goebel, J. Siekmann, and W. V

Subseries of Lecture Notes in Computer Science

John-Jules Ch. Meyer   Jan Broersen (Eds.)

Knowledge Representation for Agents and Multi-Agent Systems

First International Workshop, KRAMAS 2008
Sydney, Australia, September 17, 2008
Revised Selected Papers

 Springer

Series Editors

Randy Goebel, University of Alberta, Edmonton, Canada
Jörg Siekmann, University of Saarland, Saarbrücken, Germany
Wolfgang Wahlster, DFKI and University of Saarland, Saarbrücken, Germany

Volume Editors

John-Jules Ch. Meyer
Jan Broersen
Universiteit Utrecht, Department of Information and Computing Sciences
Padualaan 14, De Uithof, 3584 CH Utrecht, The Netherlands
E-mail: {jj,broersen}@cs.uu.nl

Library of Congress Control Number: 2009937292

CR Subject Classification (1998): I.2.4-6, I.2, F.4.1, H.3, H.2.8, F.1

LNCS Sublibrary: SL 7 – Artificial Intelligence

ISSN   0302-9743
ISBN-10   3-642-05300-9 Springer Berlin Heidelberg New York
ISBN-13   978-3-642-05300-9 Springer Berlin Heidelberg New York

springer.com

© Springer-Verlag Berlin Heidelberg 2009
Printed in Germany

Typesetting: Camera-ready by author, data conversion by Scientific Publishing Services, Chennai, India
Printed on acid-free paper   SPIN: 12780396   06/3180   5 4 3 2 1 0

Preface

This book comprises the formal proceedings of KRAMAS 2008, a workshop on Knowledge Representation for Agents and Multi-Agent Systems, held at KR 2008, Sydney, Australia, September 17, 2008. The initiative for KRAMAS 2008 was taken by last year's KR Chairs to enhance cross-fertilization between the KR (Knowledge Representation and Reasoning) and agent communities. To promote participation in the KR schedule, the workshop was conveniently 'sandwiched' between days with regular KR sessions. The topics solicited included:

- Knowledge representation and reasoning aspects of agent systems (languages, ontologies, techniques)
- Reasoning about (actions of) agents
- Reasoning methods (such as non-monotonic reasoning, abduction, argumentation, diagnosis, planning, decision making under uncertainty, reasoning about preference, ...) applied to agents and multi-agent systems (MAS)
- Theory of negotiation, communication, cooperation, group decision making, game theory for MAS
- Cognitive robotics
- Representations of other agents / opponent models
- Logics for intelligent agents and MAS
- Specification and verification techniques for agents
- Automated reasoning techniques for agent-based systems
- Logical foundations of agent-based systems, normative MAS and e-institutions
- Formal semantics of agent programming languages
- Formal techniques for agent-oriented programming and agent-oriented software engineering

We originally received 14 papers. There were two review rounds: the first one deciding on acceptance for presentation at the workshop, and the second one deciding on inclusion of revised, extended and resubmitted versions of the presented papers in these proceedings. Of the original 14, 10 papers made it into these proceedings. The workshop was a success and proved that there is indeed much interest in the problems and issues arising at the junction of KR and MAS.

We are grateful to the participants of KRAMAS 2008 and to the authors who submitted papers, to the members of the Program Committee for their service in reviewing papers (twice) and to the KR organization for taking the initiative to have KRAMAS and their support in its organization. Thanks, too, to Richard van de Stadt whose CyberChairPRO system was a very great help to us. Finally we are indebted to Springer, and Alfred Hofmann in particular, for their support in getting these proceedings published.

April 2009

John-Jules Ch. Meyer
Jan Broersen

Workshop Organization

Program Chair

John-Jules Meyer

Workshop Chairs

John-Jules Meyer   Utrecht University, The Netherlands
Jan Broersen   Utrecht University, The Netherlands

Program Committee

Thomas Agotnes	Bergen, Norway
Natasha Alechina	Nottingham, UK
Jamal Bentahar	Montreal, Canada
Rafael Bordini	Durham, UK
Jan Broersen	Utrecht, The Netherlands
Mehdi Dastani	Utrecht, The Netherlands
Giuseppe De Giacomo	Rome, Italy
Hans van Ditmarsch	Otago, New Zealand
Jurgen Dix	Clausthal, Germany
Andreas Herzig	Toulouse, France
Wiebe van der Hoek	Liverpool, UK
Wojciech Jamroga	Clausthal, Germany
Catholijn Jonker	Delft, The Netherlands
Yves Lesperance	Toronto, Canada
Alessio Lomuscio	London, UK
Timothy Norman	Aberdeen, UK
Henry Prakken	Utrecht, The Netherlands
Alessandro Ricci	Cesena, Italy
Renate Schmidt	Manchester, UK
Carles Sierra	Barcelona, Spain
Francesca Toni	London, UK
Rineke Verbrugge	Groningen, The Netherlands

Table of Contents

Reasoning about Other Agents' Beliefs under Bounded Resources

Natasha Alechina, Brian Logan, Hoang Nga Nguyen, and Abdur Rakib*

School of Computer Science
University of Nottingham
Nottingham NG8 1BB, UK
{nza,bsl,hnn,rza}@cs.nott.ac.uk

Abstract. There exists a considerable body of work on epistemic logics for bounded reasoners where the bound can be time, memory, or the amount of information the reasoners can exchange. In much of this work the epistemic logic is used as a meta-logic to reason about beliefs of the bounded reasoners from an external perspective. In this paper, we present a formal model of a system of bounded reasoners which reason about each other's beliefs, and propose a sound and complete logic in which such reasoning can be expressed. Our formalisation highlights a problem of incorrect belief ascription in resource-bounded reasoning about beliefs, and we propose a possible solution to this problem, namely adding reasoning strategies to the logic.

1 Introduction

The purpose of this paper is to investigate a multi-agent epistemic logic which results from taking seriously the idea that agents have bounded time, memory and communication resources, and are reasoning about each other's beliefs. The main contribution of the paper is to generalise several existing epistemic logics for resource-bounded reasoners by adding an ability for reasoners to reason about each other's beliefs. We show that a problem of incorrect belief ascription arises as a result, and propose a possible solution to this problem.

To give the reader an idea where the current proposal fits into the existing body of research on epistemic logics for bounded reasoners, we include a brief survey of existing approaches, concentrating mostly on the approaches which have influenced the work presented here.

In standard epistemic logic (see e.g. [1,2] for a survey) an agent's (implicit) knowledge is modelled as closed under logical consequence. This can clearly pose a problem when using an epistemic logic to model resource-bounded reasoners, whose set of beliefs is not generally closed with respect to their reasoning rules. Various proposals to modify possible worlds semantics in order to solve this problem of logical omniscience (e.g., introducing impossible worlds as in [3,4], or non-classical assignment as in [5]) result in agent's beliefs still being logically closed, but with respect to a weaker logic.

---

* This work was supported by the UK Engineering and Physical Sciences Research Council [grant number EP/E031226].

J.-J.Ch. Meyer and J.M. Broersen (Eds.): KRAMAS 2008, LNAI 5605, pp. 1–15, 2009.

Our work builds on another approach to solving this problem, namely treating beliefs as syntactic objects rather than propositions (sets of possible worlds). In [6], Fagin and Halpern proposed a model of limited reasoning using the notion of awareness: an agent explicitly believes only the formulas which are in a syntactically defined awareness set (as well as in the set of its implicit beliefs). Implicit beliefs are still closed under consequence, but explicit beliefs are not, since a consequence of explicit beliefs is not guaranteed to belong to the awareness set. However, the awareness model does not give any insight into the connection between the agent's awareness set and the agent's resource limitations, which is what we try to do in this paper.[1] Konolige [7] proposed a different model of non-omniscient reasoners, the deduction model of belief. Reasoners were parameterised with sets of rules which could, for example, be incomplete. However, the deduction model of belief still models beliefs of a reasoner as closed with respect to reasoner's deduction rules; it does not take into account the time it takes to produce this closure, or any limitations on the agent's memory. Step logic, introduced in [8], gives a syntactic account of beliefs as theories indexed by time points; each application of inference rules takes a unit of time. No fixed bound on memory was considered, but the issue of bounded memory was taken into account. An account of epistemic logic called algorithmic knowledge, which treats explicit knowledge as something which has to be computed by an agent, was introduced in [9], and further developed in e.g. [1,10]. In the algorithmic knowledge approach, agents are assumed to possess a procedure which they use to produce knowledge. In later work [10] this procedure is assumed to be given as a set of rewrite rules which are applied to the agent's knowledge to produce a closed set, so, like Konolige's approach, algorithmic knowledge is concerned with the result rather than the process of producing knowledge. In [11,12] Duc proposed logics for non-omniscient epistemic reasoners which will believe all consequences of their beliefs *eventually*, after some interval of time. It was shown in [13] that Duc's system is complete with respect to semantics in which the set of agent's beliefs is always finite. Duc's system did not model the agents' reasoning about each others' beliefs. Other relevant approaches where epistemic logics were given a temporal dimension and each reasoning step took a unit of time are, for example, [14], where each inference step is modelled as an action in the style of dynamic logic, and [15] which proposes a logic for verification of response-time properties of a system of communicating rule-based agents (each rule firing or communication takes a unit of time). In a somewhat different direction, [16] proposed a logic where agents reason about each others beliefs, but have no explicit time or memory limit; however there is a restriction on the depth of belief nestings (context switching by the agents). Epistemic logics for bounded-memory agents were investigated in, for example, [17,18,19,20], and the interplay between bounded recall and bounded memory (ability to store strategies of only bounded size) was studied in [21].

An epistemic logic BMCL for communicating agents with communication limits on the number of exchanged messages (and connections to space complexity of proofs and communication complexity) was investigated in [20]. In this paper we expand BMCL by adding rules for reasoning about other agents' beliefs, demonstrate that epistemic reasoning done in resource-bounded fashion has an inherent problem of incorrect belief ascription, and propose the use of reasoning strategies as a solution to this problem.

---

[1] We also completely dispense with the notion of implicit beliefs.

2   Model of Reasoning Agents

The logic BMCL presented in [20] formalises reasoning about the beliefs of a system of reasoners who reason using propositional resolution and can exchange information to solve a problem together. The set up is similar to, for example, [22]. BMCL models each inference rule application as taking a single time step, introduces an explicit bound on the set of beliefs of each reasoner, and a bound on the number of messages the reasoners can exchange. In this paper, we generalise this approach by assuming that agents can also reason about each other's beliefs. Namely, they assume that other agents use a certain set of inference rules, and they reason about what another agent may believe at the next step. For example, if agent A believes that agent B believes two clauses c_1 and c_2 and these two clauses are resolvable to a clause c, and agent A assumes that agent B reasons using resolution, then it is reasonable for agent A to believe that agent B may believe c at the next step.

We assume a set of n agents. Each agent i has a set of inference rules, a set of premises KB_i, and a *working memory*. To infer from the premises in KB_i, the relevant formulas must first be read into working memory. We assume that each agent's working memory is bounded by n_M, which is the maximal number of formulas an agent can believe at the same time. We also set a limit on the possible size of a formula, or rather on the depth of nesting of belief operators, n_B, and a limit, n_C, on the maximal number of communications an agent can make. For simplicity, we assume that these bounds are the same for all agents, but this can be easily relaxed by introducing functions $n_M(i)$, $n_B(i)$ and $n_C(i)$ which assign a different limit to each agent i.

The set of reasoning actions is as follows:

Read KB: an agent can retrieve information from its KB and put it into its working memory using the *Read* action. Since an agent has a fixed size memory, adding a formula to its memory may require erasing some belief already in memory (if the limit n_M would otherwise be exceed). The same applies to other reasoning actions which add a new formula, in that adding a new formula may involve overwriting a formula currently in working memory.

Resolution: an agent can derive a new clause if it has two resolvable clauses in its memory.

Copy: an agent can communicate with another agent to request a clause from the memory of the other agent. We assume that communication is always successful if the other agent has the requested clause. If agent A has clause c in memory, then a copy by B will result in agent B believing that A believes c. *Copy* is only enabled if the agent has performed fewer than n_C copy actions in the past and the prefix of the resulting belief has nesting of at most n_B.

Idle: an agent may idle (do nothing) at any time step. This means that at the next time point of the system, the agent does not change its state of memory.

Erase: an agent may remove a formula from its working memory. This action is introduced for technical reasons to simplify the proofs.

In addition to the actions listed above, we introduce actions that enable agents to reason about other agents' beliefs, essentially epistemic axioms K (ascribing propositional reasoning to the other agent) and 4 (positive introspection about the agent's own beliefs, and ascribing positive introspection to other agents). The reasons we do not adopt for example KD45 are as follows. If the agent's knowledge base is inconsistent, we want it to be able to derive $B\bot$ (or $B[]$ where $[]$ is the empty clause). Negative introspection is also problematic in a resource-bounded setting, in that the agent may derive $\neg B\alpha$ if α is not in its current set of beliefs, and then derive α from its other beliefs, ending up with an inconsistent set of beliefs ($\neg B\alpha$ and $B\alpha$ by positive introspection from α), even if its knowledge base is consistent. We could have adopted a restricted version of negative introspection (see, e.g., [12]) but in this paper we omit it for simplicity.

In addition to the reasoning actions listed above, we therefore add the following actions:

Other's Resolution: an agent A can perform this action if it believes that another agent B believes two resolvable clauses c_1 and c_2. Then A can conclude that B will believe in the resolvent clause c of c_1 and c_2 in the next time point. As a general case, we can extend the chain *agent-believes ... agent-believes*. For example, if agent A believes that agent B believes that agent C believes two resolvable clauses c_1 and c_2, then it is possible in the next time point that agent A believes that agent B believes that agent C believes c which is the resolvent of c_1 and c_2.

Positive Introspection: if an agent A believes a clause c, it can perform this action to reach a state where it believes that it believes c.

Other's Positive Introspection: if an agent A believes that another agent B believes a clause c, it can perform this action to reach a state where it believes that B believes that B believes c.

The reasoning actions *Positive Introspection* and *Other's Positive Introspection* are only enabled if the derived formula has a depth of nesting of at most n_B.

Note that the assumption that the agents reason using resolution and positive introspection is not essential for the main argument of this paper. This particular set of inference rules has been chosen to make the logic concrete; we could have, for example, assumed that the agents reason using modus ponens and conjunction introduction instead of resolution. In what follows, we give a formal definition of an epistemic logic for communicating agents which reason in a step-wise, memory-bounded fashion using some well-defined set of inference rules.

3   Syntax and Semantics of $ERBL$

In this section, we give the syntax and semantics of the logic $ERBL$ which formalises the ideas sketched in the previous section. ERBL (Epistemic Resource Bounded Logic) is an epistemic and temporal meta-language in which we can talk about beliefs expressed in the agents' internal language.

Let the set of agents be $A = \{1, 2, \ldots, n_A\}$. We assume that all agents agree on a finite set $PROP$ of propositional variables, and that all *belief formulas* of the internal language of the agents are in the form of *clauses* or clauses preceded by a prefix of belief operators of fixed length.

From the set of propositional variables, we have the definition of all literals as follows:

$$LPROP = \{p, \neg p \mid p \in PROP\}$$

Then, the set of all clauses is $\Omega = \wp(LPROP)$. Finally, the set of all belief formulas is defined as follows:

$$B\Omega ::= \{B_{i_1} \ldots B_{i_k} c \mid c \in \Omega, 0 \leq k \leq n_B\},$$

where $i_j \in A$. Note that we only include in the set of belief formulas those whose belief operator nesting is limited by n_B. Therefore, $B\Omega$ is finite.

Each agent $i \in A$ is assumed to have a knowledge base $KB_i \subseteq B\Omega$.

For convenience, the negation of a literal L is defined as $\neg L$, where:

$$\neg L = \begin{cases} \neg p & \text{if } L = p \text{ for some } p \in PROP \\ p & \text{if } L = \neg p \text{ for some } p \in PROP \end{cases}$$

The form of resolution rule which will be used in formal definitions below is as follows: given two clauses c_1 and $c_2 \in \Omega$ such that one contains a literal L and the other has its negation $\neg L$, we can derive a new clause which is the union $c_1 \setminus \{L\} \cup c_2 \setminus \{\neg L\}$.

The syntax of $ERBL$ is then defined inductively as follows.

- $\top$ is a well-formed formula (wff) of $ERBL$.
- $start$ is a wff of $ERBL$; it is a marker for the start state.
- $cp_i^{=n}$ (the number of communication actions performed by agent i) is a wff of $ERBL$ for all $n = 0, \ldots, n_C$, and $i \in A$; it is used as a communication counter in the language.
- If $\alpha \in B\Omega$, then $B_i\alpha$ (agent i believes α) is a wff of $ERBL$, $i \in A$.
- If φ and ψ are wffs, then so are $\neg\varphi$, $\varphi \wedge \psi$.
- If φ and ψ are wffs, then so are $X\varphi$ (φ holds in the next moment of time), $\varphi U \psi$ (φ holds until ψ), and $A\varphi$ (φ holds on all paths).

Classical abbreviations for $\vee$, $\rightarrow$, $\leftrightarrow$ are defined as usual. We also have $\bot \equiv \neg\top$, $F\varphi \equiv \top U \varphi$ (φ holds some time in the future), $E\varphi \equiv \neg A \neg \varphi$ (φ holds on some path). For convenience, let $CP_i = \{cp_i^{=n} \mid n = \{0, \ldots, n_C\}\}$ and $CP = \bigcup_{i \in A} CP_i$.

The semantics of $ERBL$ is defined by $ERBL$ transition systems which are based on ω-tree structures (standard CTL^* models as defined in [23]).

Let (T, R) be a pair where T is a set and R is a binary relation on T. Let the relation $<$ be the irreflexive and transitive closure of R, namely the set of pairs of states $\{(s, t) \in T \times T \mid \exists n \geq 0, t_0 = s, .., t_n = t \in T$ such that $t_i R t_{i+1}$ for all $i = 0, \ldots, n-1\}$. (T, R) is a ω-tree frame iff the following conditions are satisfied.

1. T is a non-empty set.
2. R is total, i.e., for all $t \in T$, there exists $s \in T$ such that tRs.
3. For all $t \in T$, the past $\{s \in T \mid s < t\}$ is linearly ordered by $<$.
4. There is a smallest element called the root, denoted by t_0.
5. Each maximal linearly $<$- ordered subset of T is order-isomorphic to the natural numbers.

A branch of (T, R) is an ω-sequence $(t_0, t_1, \ldots)$ such that t_0 is the root and $t_i R t_{i+1}$ for all $i \geq 0$. We denote by $B(T, R)$ the set of all branches of (T, R).

A $ERBL$ transition system M is defined as a triple (T, R, V) where:

- (T, R) is a ω-tree frame,
- $V : T \times A \rightarrow \wp(B\Omega \cup CP)$ such that for all $s \in T$ and $i \in A$: $V(s, i) = Q \cup \{cp_i^{=n}\}$ for some $Q \subseteq B\Omega$ and $0 \leq n \leq n_C$. We denote by $V^*(s, i)$ the set $V(s, i) \setminus \{cp_i^{=n} | 0 \leq n\}$.

For a branch $\sigma \in B(T, R)$, σ_i denotes the element t_i of σ and $\sigma_{\leq i}$ is the prefix $(t_0, t_1, \ldots, t_i)$ of σ.

The truth of a $ERBL$ formula at a point n of a path $\sigma \in B(T, R)$ is defined inductively as follows:

- $M, \sigma, n \models \top$,
- $M, \sigma, n \models B_i \alpha$ iff $\alpha \in V(s, i)$,
- $M, \sigma, n \models start$ iff $n = 0$,
- $M, \sigma, n \models cp_i^{=m}$ iff $cp_i^{=m} \in V(s, i)$,
- $M, \sigma, n \models \neg \varphi$ iff $M, \sigma, n \not\models \varphi$,
- $M, \sigma, n \models \varphi \wedge \psi$ iff $M, \sigma, n \models \varphi$ and $M, \sigma, n \models \psi$,
- $M, \sigma, n \models X\varphi$ iff $M, \sigma, n + 1 \models \varphi$,
- $M, \sigma, n \models \varphi U \psi$ iff $\exists m \geq n$ such that $\forall k \in [n, m)$ $M, \sigma, k \models \varphi$ and $M, \sigma, m \models \psi$,
- $M, \sigma, n \models A\varphi$ iff $\forall \sigma' \in BR$ such that $\sigma'_{\leq n} = \sigma_{\leq n}$, $M, \sigma', n \models \varphi$.

The set of possible transitions in a model is defined as follows. Definition 1 below describes possible outcomes of various actions. For example, performing a resolution results in adding the resolvent to the set of beliefs. Definition 2 describes when an action is possible or enabled. For example, resolution is enabled if the agent has two resolvable clauses in memory.

Definition 1. *Let (T, R, V) be a tree model. The set of effective transitions R_a for an action a is defined as a subset of R and satisfies the following conditions, for all $(s, t) \in R$:*

1. $(s, t) \in R_{Read_{i, \alpha, \beta}}$ *iff* $\alpha \in KB_i$, $\alpha \notin V(s, i)$ *and* $V(t, i) = V(s, i) \setminus \{\beta\} \cup \{\alpha\}$. This condition says that s and t are connected by agent i's $Read$ transition if the following is true: α is in i's knowledge base but not in $V(s, i)$, α is added to the set of i's beliefs at t, and $\beta \in B\Omega$ is removed from the agent's set of beliefs. The argument β stands for a formula which is overwritten in the transition. If $\beta \in V(s, i)$ then the agent actually loses a belief in the transition, if $\beta \notin V(s, i)$ then the transition only involves adding a formula α without removing any beliefs.

2. $(s, t) \in R_{Res_{i, \alpha_1, \alpha_2, L, \beta}}$ *where* $\alpha_1 = B_{i_1} \ldots B_{i_{k-1}} B_{i_k} c_1$ *and* $\alpha_2 = B_{i_1} \ldots B_{i_{k-1}} B_{i_k} c_2$ *iff* $\alpha_1 \in V(s, i)$, $\alpha_2 \in V(s, i)$, $L \in c_1$, $\neg L \in c_2$, $\alpha = B_{i_1} \ldots B_{i_{k-1}} B_{i_k} c \notin V(s, i)$ *and* $V(t, i) = V(s, i) \setminus \{\beta\} \cup \{\alpha\}$ *where* $c = c_1 \setminus \{L\} \cup c_2 \setminus \{\neg L\}$. This condition says that s and t are connected by agent i's Res transition if in s agent i believes two resolvable clauses α_1 and α_2 but not α, possibly preceded by the same sequence of belief operators, and in t agent i believes their resolvent, preceded by the same prefix. Again, $\beta \in B\Omega$ is overwritten if it is in the set of agent's beliefs in s.

3. $(s,t) \in R_{Copy_{i,\alpha,\beta}}$ iff $\alpha \in V(s,j)$ for some $j \in A$ and $j \neq i$, for any $cp_i^{=n} \in V(s,i)$ such that $n < n_C$, $B_j\alpha \notin V(s,i)$ and $V(t,i) = V(s,i) \setminus \{cp_i^{=n} | cp_i^{=n} \in V(s,i)\} \cup \{cp_i^{=n+1} | cp_i^{=n} \in V(s,i)\} \setminus \{\beta\} \cup \{B_j\alpha\}$. s and t are connected by a *Copy* transition of agent i if in t, i adds to its beliefs a formula $B_j\alpha$ where α is an agent j's belief in s, and i has previously copied fewer than n_C formulas. Again some $\beta \in B\Omega$ is possibly overwritten.

4. $(s,t) \in R_{Idle_i}$ iff $V(t,i) = V(s,i)$. The *Idle* transition does not change the state.

5. $(s,t) \in R_{Erase_{i,\beta}}$ iff $V(t,i) = V(s,i) \setminus \{\beta\}$. *Erase* removes one of the agent's beliefs.

6. $(s,t) \in R_{PI_{i,\alpha,\beta}}$ iff $\alpha \in V(s,i)$, $B_i\alpha \notin V(s,i)$ and $V(t,i) = V(s,i) \setminus \{\beta\} \cup \{B_i\alpha\}$. *PI* is i's positive introspection: s and t are connected by i's *PI* transition if in s it believes α but not $B_i\alpha$ and in t it believes $B_i\alpha$.

7. $(s,t) \in R_{OPI_{i,B_{i_1} \ldots B_{i_{k-1}} B_{i_k}\alpha,\beta}}$ iff $B_{i_1} \ldots B_{i_{k-1}} B_{i_k}\alpha \in V(s,i)$ but not $B_{i_1} \ldots B_{i_{k-1}} B_{i_k} B_{i_k}\alpha$, $V(t,i) = V(s,i) \setminus \{\beta\} \cup \{B_{i_1} \ldots B_{i_{k-1}} B_{i_k} B_{i_k}\alpha\}$. This corresponds to ascribing positive introspection to agent i_k.

This specifies the effects of actions. Below, we specify when an action is possible. Note that we only enable deriving a formula if this formula is not already in the set of the agent's beliefs.

Definition 2. *Let (T, R, V) be a tree model. The set $Act_{s,i}$ of possible actions that an agent i can perform at a state $s \in T$ is defined as follows:*

1. $Read_{i,\alpha,\beta} \in Act_{s,i}$ iff $\alpha \notin V(s,i)$, $\alpha \in KB_i$ and $\beta \in V(s,i)$ if $|V^*(s,i)| \geq n_M$.
2. $Res_{i,\alpha_1,\alpha_2,L,\beta} \in Act_{s,i}$ iff $c = (c_1 \setminus L) \cup (c_2 \setminus \neg L) \notin V(s,i)$, $\alpha_1 = B_{i_1} \ldots B_{i_{k-1}} B_{i_k} c_1$, $\alpha_2 = B_{i_1} \ldots B_{i_{k-1}} B_{i_k} c_2$, $L \in c_1$, $\neg L \in c_2$, $\alpha_1, \alpha_2 \in V(s,i)$, and $\beta \in V(s,i)$ if $|V^*(s,i)| \geq n_M$.
3. $Copy_{i,\alpha,\beta} \in Act_{s,i}$ iff $B_j\alpha \notin V(s,i)$, $\alpha \in V(s,j)$ for some $j \in A$ and $j \neq i$, $n < n_C$ for any $cp_i^{=n} \in V(s,i)$ and $\beta \in V(s,i)$ if $|V^*(s,i)| \geq n_M$.
4. It is always the case that $Idle_i \in Act_{s,i}$.
5. $PI_{i,\alpha,\beta} \in Act_{s,i}$ iff $B_i\alpha \notin V(s,i)$, $\alpha \in V(s,i)$ and $\beta \in V(s,i)$ if $|V^*(s,i)| \geq n_M$.
6. $OPI_{i,B_{i_1} \ldots B_{i_{k-1}} B_{i_k}\alpha,\beta} \in Act_{s,i}$ iff $B_{i_1} \ldots B_{i_{k-1}} B_{i_k} B_{i_k}\alpha \notin V(s,i)$, $B_{i_1} \ldots B_{i_{k-1}} B_{i_k}\alpha \in V(s,i)$ and $\beta \in V(s,i)$ if $|V^*(s,i)| \geq n_M$.

There are no specified conditions for enabling $Erase_{i,\beta}$. This action is introduced for technical reasons, to simplify the proofs.

Finally, the definition of the set of models corresponding to a system of reasoners is given below:

Definition 3. $M(KB_1, \ldots, KB_{n_A}, n_B, n_M, n_C)$ *is the set of models (T, R, V) which satisfies the following conditions:*

1. $|V^*(s,i)| \leq n_M$ for all $s \in T$ and $i \in A$.
2. $cp_i^{=0} \in V(t_0, i)$ where t_0 is the root of (T, R) for all $i \in A$.
3. $R = \bigcup_{\forall a} R_a$.
4. For all $s \in T$, $a_i \in Act_{s,i}$, there exists $t \in T$ such that $(s,t) \in R_{a_i}$ for all $i \in A$.

4 Axiomatisation

In this section, we introduce an axiom system which is sound and complete with respect to the set of models defined in the previous section.

Below are some abbreviations which will be used in the axiomatisation:

- $ByRead_i(\alpha, n) = \neg B_i\alpha \wedge cp_i^{=n}$. This formula describes the state before the agent comes to believe formula α by the $Read$ transition. n is the value of i's communication counter.
- $ByRes_i(\alpha, n) = \bot$ if $\alpha = B_{i_1} \ldots B_{i_{k-1}} \neg B_{i_k} c$ for some $c \in \Omega$ and $1 \leq k \leq n_B$; otherwise $ByRes_i(\alpha, n) = \neg B_i\alpha \wedge \bigvee_{(\alpha_1, \alpha_2) \in Res^{-1}(\alpha)} (B_i\alpha_1 \wedge B_i\alpha_2)$ where $Res^{-1}(B_{i_1} \ldots B_{i_{k-1}} B_{i_k} c) = \{(B_{i_1} \ldots B_{i_{k-1}} B_{i_k} c_1, B_{i_1} \ldots B_{i_{k-1}} B_{i_k} c_2) \mid \exists L \in LPROP$ such that $c = c_1 \setminus \{L\} \cup c_2 \setminus \{\neg L\}\}$. This formula describes the state of the system before i derives α by resolution. Note that it may not be possible to derive an arbitrary formula α by resolution; in that case, the state is described by falsum $\bot$.
- $ByCopy_i(\alpha, n) = \bot$ if $\alpha \neq B_j\alpha'$ for some $j \neq i$ or $n \leq 0$; otherwise $ByCopy_i(B_j\alpha', n) = \neg B_i B_j\alpha' \wedge B_j\alpha' \wedge cp^{=n-1}$.
- $ByPI_i(\alpha, n) = \bot$ if $\alpha \neq B_i\alpha'$; otherwise $ByPI_i(\alpha, n) = \neg B_i B_i\alpha' \wedge B_i\alpha' \wedge cp^{=n}$.
- $ByOPI_i(\alpha, n) = \bot$ if $\alpha \neq B_{i_1} \ldots B_{i_{k-1}} B_{i_k} B_{i_k}\alpha'$; otherwise $ByOPI_i(\alpha, n) = \neg B_i B_{i_1} \ldots B_{i_{k-1}} B_{i_k} B_{i_k}\alpha' \wedge B_i B_{i_1} \ldots B_{i_{k-1}} B_{i_k}\alpha' \wedge cp^{=n}$.

The axiomatisation is as follows.

A1. All axioms and inference rules of CTL^* [24].

A2. $\bigwedge_{\gamma \in Q} B_i\gamma \wedge cp_i^{=n} \wedge \neg B_i\alpha \rightarrow EX(\bigwedge_{\gamma \in Q} B_i\gamma \wedge cp_i^{=n} \wedge B_i\alpha)$ for all $\alpha \in KB_i$, and $Q \subseteq B\Omega$ such that $|Q| < n_M$.

Intuitively, this axiom says that it is always possible to make a transition to a state where agent i believes a formula from its knowledge base KB_i. In addition, the communication counter of the agent does not increase, and a set of beliefs Q of cardinality less than n_M can also be carried over to the same state.

Axioms **A3** - **A6** similarly describe transitions made by resolution (given that resolvable clauses are available), copy (with communication counter increased), and positive introspection (applied by agent i or ascribed by i to another agent).

A3. $\bigwedge_{\gamma \in Q} B_i\gamma \wedge B_i B_{i_1} \ldots B_{i_{k-1}} B_{i_k} c_1 \wedge B_i B_{i_1} \ldots B_{i_{k-1}} B_{i_k} c_2 \wedge cp_i^{=n} \wedge \neg B_i B_{i_1} \ldots B_{i_{k-1}} B_{i_k} c \rightarrow EX(\bigwedge_{\gamma \in Q} B_i\gamma \wedge cp_i^{=n} \wedge B_i B_{i_1} \ldots B_{i_{k-1}} B_{i_k} c)$ for all $c_1, c_2 \in \Omega$ such that $L \in c_1$, $\neg L \in c_2$ and $c = c_1 \setminus \{L\} \cup c_2 \setminus \{\neg L\}$, $k \geq 0$, and $Q \subseteq B\Omega$ such that $|Q| < n_M$.

A4. $\bigwedge_{\gamma \in Q} B_i\gamma \wedge B_j\alpha \wedge cp_i^{=n} \wedge \neg B_i B_j\alpha \rightarrow EX(\bigwedge_{\gamma \in Q} B_i\gamma \wedge B_i B_j\alpha \wedge cp_i^{=n+1})$ for any $\alpha \in B\Omega$, $j \in A$, $j \neq i$, $n < n_C$, and $Q \subseteq B\Omega$ such that $|Q| < n_M$.

A5. $\bigwedge_{\gamma \in Q} B_i\gamma \wedge B_i\alpha \wedge cp_i^{=n} \wedge \neg B_i B_i\alpha \rightarrow EX(\bigwedge_{\gamma \in Q} B_i\gamma \wedge B_i B_i\alpha \wedge cp_i^{=n})$ for any $\alpha \in B\Omega$ and $Q \subseteq B\Omega$ such that $|Q| < n_M$.

A6. $\bigwedge_{\gamma \in Q} B_i\gamma \wedge B_i B_{i_1} \ldots B_{i_{k-1}} B_{i_k}\alpha \wedge cp_i^{=n} \wedge \neg B_i B_{i_1} \ldots B_{i_{k-1}} B_{i_k} B_{i_k}\alpha \rightarrow EX(\bigwedge_{\gamma \in Q} B_i\gamma \wedge B_i B_{i_1} \ldots B_{i_{k-1}} B_{i_k} B_{i_k}\alpha \wedge cp_i^{=n})$ for any $\alpha \in B\Omega$, $k \geq 0$ and $Q \subseteq B\Omega$ such that $|Q| < n_M$.

A7. $EX(B_i\alpha \wedge B_i\beta) \rightarrow B_i\alpha \vee B_i\beta$.

This axiom says that at most one new belief is added in the next state.

A8. $EX(\neg B_i\alpha \wedge \neg B_i\beta) \rightarrow \neg B_i\alpha \vee \neg B_i\beta$.

This axiom says that at most one belief is deleted in the next state.

A9. $EX(B_i\alpha \wedge cp_i^{=n}) \rightarrow B_i\alpha \vee ByRead_i(\alpha, n) \vee ByRes_i(\alpha, n) \vee ByCopy_i(\alpha, n) \vee ByPI_i(\alpha, n) \vee ByOPI_i(\alpha, n)$ for $\alpha \in KB_i$.

This axiom says that a new belief which is an element of the agent's knowledge base can only be added by one of the valid reasoning actions.

A10. $EX(B_i\alpha \wedge cp_i^{=n}) \rightarrow B_i\alpha \vee ByRes_i(\alpha, n) \vee ByCopy_i(\alpha, n) \vee ByPI_i(\alpha, n) \vee ByOPI_i(\alpha, n)$ for $\alpha \notin KB_i$.

This axiom describes possible ways in which a new belief which is not in the agent's knowledge base can be added.

A11. $B_i\alpha_1 \wedge \ldots \wedge B_i\alpha_{n_M} \rightarrow \neg B_i\alpha_{n_M+1}$ for all $i \in A$, $\alpha_j \in B\Omega$ where $j = 1, \ldots, n_M + 1$ and all α_j are pairwise different.

This axiom states that an agent cannot have more than n_M different beliefs.

A12a $start \rightarrow cp_i^{=0}$ for all $i \in A$.

In the start state, the agent has not performed any $Copy$ actions.

A12b $\neg EX start$ ($start$ only holds at the root of the tree).

A13. $\bigvee_{n=0 \ldots n_C} cp_i^{=n}$ for all $i \in A$.

There is always a number n between 0 and n_C corresponding to the number of $Copy$ actions agent i has performed.

A14. $cp_i^{=n} \rightarrow \neg cp_i^{=n'}$ for all $i \in A$ and $n' \neq n$.

The number of previous $Copy$ actions by i in each state is unique.

A15. $\varphi \rightarrow EX\varphi$ where φ does not contain $start$.

It is always possible to make a transition to a state where all agents have the same beliefs and communication counter values as in the current state (essentially an $Idle$ transition by all agents).

A16. $\bigwedge_{i \in A} EX(\bigwedge_{\gamma \in Q_i} B_i\gamma \wedge cp_i^{=n_i}) \rightarrow EX \bigwedge_{i \in A}(\bigwedge_{\gamma \in Q_i} B_i\gamma \wedge cp_i^{=n_i})$ for any $Q_i \subseteq B\Omega$ such that $|Q_i| \leq n_M$.

If each agent i can separately reach a state where it believes formulas in Q_i, then all agents together can reach a state where for each i, agent i believes formulas in Q_i.

Notice that since the depth of the nesting of belief operators is restricted by n_B, for any subformula $B_i\alpha$ appearing in any above axiom, $\alpha \in B\Omega$.

Definition 4. $L(KB_1, \ldots, KB_{n_A}, n_B, n_M, n_C)$ is the logic defined by the axiomatisation *A1–A16*.

We have the following result.

Theorem 1. $L(KB_1, \ldots, KB_{n_A}, n_B, n_M, n_C)$ *is sound and complete with respect to* $M(KB_1, \ldots, KB_{n_A}, n_B, n_M, n_C)$.

The proof is omitted due to lack of space; it is based on the proof technique used in [24].

5   Discussion

Systems of step-wise reasoners with bounded memory and a communication limit are faithful models of systems of distributed resource-limited reasoners, and various resource requirements of such systems can be effectively verified, e.g. by model-checking, as in for example [20]. However, adding reasoning about beliefs poses a significant challenge, both in the complexity of the system and in the way this reasoning is modelled. The branching factor of the models is much larger when reasoning about beliefs is considered, making model-checking less feasible. The main problem however has to do with the correctness of an agent's belief ascription. We describe this problem below and propose a tentative solution.

In the system proposed in this paper, agents correctly ascribe reasoning mechanisms to each other, and in the limit, their predictions concerning other agents' beliefs are correct: if agent j believes that eventually agent i will believe α, then eventually agent i will believe α, and vice versa. More precisely, for every model M and state s,

$$\{\alpha : M, s \models EFB_jB_i\alpha\} = \{\alpha : M, s \models EFB_i\alpha\}$$

However, in spite of this, the agents are almost bound to make wrong predictions when trying to second-guess what other reasoners will believe in the next state. More precisely,

$$\{\alpha : M, s \models B_jB_i\alpha\} \not\subseteq \{\alpha : M, s \models B_i\alpha\}$$

i.e. agent j may believe that i believes some α when i does not believe α.

Consider the following example. Suppose there are two agents, 1 and 2, each with a memory limit of two formulas, communication limit of one formula, belief nesting limit of two, and knowledge bases $KB_1 = \{p\}$ and $KB_2 = \{q\}$. A possible run of the system is shown in Figure 1.

State	Agent 1	Agent 2
t_0	{ }	{ }
transition:	Read	Read
t_1	$\{p\}$	$\{q\}$
transition:	Copy	Copy
t_2	$\{p, B_2q\}$	$\{q, B_1p\}$

Fig. 1. A possible run of the system

Note that this is only one possible run, and other transitions are possible. For example, in t_0, one or both agents can idle. In t_1, one or both agents can idle, or make a positive introspection transition. In state t_2, the agents' beliefs about each other's beliefs are correct. However, in most successor states of t_2, agent 1 will have incorrect beliefs about agent 2's beliefs, and vice versa. Indeed, the options of agent 1 in t_2 are: read p, idle, erase p, erase B_2q, apply positive introspection to derive B_1p or B_1B_2q, ascribe introspection to agent 2 to derive B_2B_2q. Agent 2 has similar choices. In only two of these cases do the agents make non-trivial (that is, new compared to the ones already existing in t_2) correct belief ascriptions, namely if agent 1 derives B_1p and agent

State	Agent 1	Agent 2
t_2	$\{p, B_2q\}$	$\{q, B_1p\}$
transition:	PI, overwrite B_2q	OPI, overwrite q
t_3	$\{p, B_1p\}$	$\{B_1p, B_1B_1p\}$

Fig. 2. Continuing from t_2: a correct ascription

2 derives B_1B_1p, and vice versa when agent 2 derives B_2q and agent 1 derives B_2B_2q (see Figure 2).

Figure 3 shows one of many possible incorrect ascriptions. Note that agent 1's ascription is now incorrect because agent 2 has forgotten q, and agent 2's ascription is incorrect because it assumed agent 1 will use positive introspection to derive B_1p, which it did not.

State	Agent 1	Agent 2
t_2	$\{p, B_2q\}$	$\{q, B_1p\}$
transition:	Idle	OPI, overwrite q
t_4	$\{p, B_2q\}$	$\{B_1p, B_1B_1p\}$

Fig. 3. Continuing from t_2: an incorrect ascription

This suggests an inherent problem with modelling agents reasoning about each other's beliefs in a step-wise, memory-bounded fashion. Note that this problem is essentially one of belief ascription, i.e., of correctly predicting what another agent will believe given limited information about what it currently believes (of deriving correct conclusions from correct premises), rather than a problem of belief revision [25], i.e., what an agent should do if it discovers the beliefs it has ascribed to another agent are incorrect. It is also distinct from the problem of determining the consequences of information updates as studied in dynamic epistemic logic (e.g. [26]). Adding new true beliefs in a syntactic approach such as ours is straightforward compared to belief update in dynamic epistemic logic, which interprets beliefs as sets of possible worlds. Essentially, in dynamic epistemic logic an agent acquires a new logically closed set of beliefs at the next 'step' after an announcement is made, while we model the gradual process of deriving consequences from a new piece of information (and the agent's previous beliefs).

The disparity between agent i's beliefs and the beliefs agent j ascribes to i at each step is due both to the fact that at most one formula is derived by each agent at any given step (and agent j may guess incorrectly which inference rule agent i is going to use) and to memory limitations which cause agents to forget formulas. An obvious alternative is to do tentative ascription of beliefs to other agents, namely conclude that the other agent will be in *one of several* possible belief sets in the next state, e.g.

$$B_2B_1p \rightarrow EX(B_2((B_1p \wedge B_1B_1p) \vee (B_1p \wedge \neg B_1B_1p) \vee \ldots))$$

However, this implies that one of the agents (agent 2 in this case) has a much larger (exponentially larger!) memory and a more expressive internal language to reason about the other agent's beliefs.

It is clearly not sufficient for correct belief prediction for the reasoners to ascribe to other agents just a set of inferences rules or a logic such as KD45. They need to be able to ascribe to other agents a *reasoning strategy*, or a preference order on the set reasoning actions used by the other agents which constrains the possible transitions of each reasoner, and directs each agent's reasoning about the beliefs of other agents. As a simple example, suppose agent 2 believes that agent 1's strategy is to apply positive introspection to formula p in preference to all other actions. Then in state t_2 agent 2 will derive B_1B_1p from B_1p. If agent 2's ascription of strategy to agent 1 is correct, agent 1 will indeed derive B_1p from p in the next state, making agent 2's belief prediction correct.

6   *ERBL* with strategies

In this section, we modify the semantics of $ERBL$ to introduce reasoning strategies.

First we need to define strategies formally. A *reasoning strategy for agent* i, $\prec_i$, is a total order on the set Act_i of all reasoning actions of i and their arguments:

$$Act_i = \{Read_{i,\alpha,\beta},\ Res_{i,\alpha_1,\alpha_2,L,\beta},\ Copy_{i,\alpha,\beta},$$
$$Erase_{i,\beta},\ Idle_i,\ PI_{i,\alpha,\beta},\ OPI_{i,B_{i_1}\ldots B_{i_{k-1}},B_{i_k}\alpha,\beta} \mid \alpha,\beta,\alpha_1,\alpha_2 \in B\Omega\}$$

A simple example of a reasoning strategy for i would be a lexicographic order on Act_i which uses two total orders: an order on the set of transitions, e.g. $Res < PI < OPI < Copy < Read < Idle$, and an order on $B\Omega$.

Recall that in Definition 2 we specified which actions are enabled in state s, $Act_{s,i} \subseteq Act_i$. We required in Definition 3 that for each enabled action, there is indeed a transition by that action out of s. The simple change that we make to Definition 3 is that for every agent i we only enable *one* action, namely the element of $Act_{s,i}$ which is minimal in $\prec_i$.

Definition 5. *The set of reasoning strategy models* $M^{strat}(KB_1,\ldots,KB_{n_A},n_B,n_M, n_C)$ *is the set of models* (T,R,V) *which satisfies conditions 1-3 from Definition 3 and the following condition:*

4'. *For all* $s \in T$, *there exists a unique state* t *such that* $(s,t) \in R_{a_i}$ *for all* $i \in A$, *where* a_i *is the minimal element with respect to* $\prec_i$ *in* $Act_{s,i}$.

Observe that in the reasoning strategy models, the transition relation is a linear order.

Finally, we give one possible definition of a correct ascription of a reasoning strategy which allows an agent j to have a correct and complete representation of the beliefs of another agent i, namely ensuring that $B_i\alpha \leftrightarrow B_jB_i\alpha$ at each step. Such perfect matching of i's beliefs by j is possible if

$$KB_j = \{B_i\alpha : \alpha \in KB_i\}$$

and agent i does not use the *Copy* action (intuitively, because in order to match *Copy* by i, agent j has to add two modalities in one step: when agent i derives $B_l\alpha$ from α being in agent l's belief set, agent j has to derive $B_iB_l\alpha$). Below, we also assume that j is allowed one extra nesting of belief modalities ($n_B(j) = n_B(i) + 1$).

Definition 6. *Agent j has a strategy which* matches *the strategy of agent i if for every natural number k, the following correspondence holds between the kth element of $\prec_j$ and the kth element of $\prec_i$:*

- *if the kth element of $\prec_i$ is $Read_{i,\alpha,\beta}$, then the kth element of $\prec_j$ is $Read_{j,B_i\alpha,B_i\beta}$*
- *if the kth element of $\prec_i$ is $Res_{i,\alpha_1,\alpha_2,L,\beta}$, then the kth element of $\prec_j$ is $Res_{j,B_i\alpha_1,B_i\alpha_2,L,B_i\beta}$*
- *if the kth element of $\prec_i$ is $PI_{i,\alpha,\beta}$, then the kth element of $\prec_j$ is $OPI_{j,B_i\alpha,B_i\beta}$*
- *if the kth element of $\prec_i$ is $OPI_{i,B_l\alpha,\beta}$, then then the kth element of $\prec_j$ is $OPI_{j,B_iB_l\alpha,B_i\beta}$.*
- *if the kth element of $\prec_i$ is $Erase_{i,\beta}$, then the kth element of $\prec_j$ is $Erase_{j,B_i\beta}$*
- *if the kth element of $\prec_i$ is $Idle_i$, then the kth element of $\prec_j$ is $Idle_j$.*

Theorem 2. *If agent j's strategy matches agent i's strategy and agent j has complete and correct beliefs about agent i's beliefs in state s: $M, s \models B_i\alpha \leftrightarrow B_jB_i\alpha$, then agent j will always have correct beliefs about agent i's beliefs: $M, s \models AG(B_i\alpha \leftrightarrow B_jB_i\alpha)$.*

Other more realistic matching strategies, for example, those which allow the agent to have a less than complete representation of other agent's beliefs, are possible, and their formal investigation is a subject of future work.

7   Conclusion

We presented a formal model of resource-bounded reasoners reasoning about each other's beliefs, and a sound and complete logic, $ERBL$, for reasoning about such systems. Our formalisation highlighted a problem of incorrect belief ascription, and we showed that this problem can be overcome by extending the framework with reasoning strategies. In future work we plan to extend the framework in a number of ways, including producing correct belief ascription under less strict matching between agents' strategies, and introducing reasoning about other agent's resource limitations. At the moment the agents have no way of forming beliefs about another agent's memory limit n_M or belief nesting bound n_B (note that we can also easily make those limits different for different agents). If they could represent those limitations, then one agent could infer that another agent does not believe some formula on the grounds that the latter agent's memory is bounded.

References

1. Fagin, R., Halpern, J.Y., Moses, Y., Vardi, M.Y.: Reasoning about Knowledge. MIT Press, Cambridge (1995)
2. Meyer, J.J., van der Hoek, W.: Epistemic Logic for Computer Science and Artificial Intelligence. Cambridge University Press, Cambridge (1995)
3. Hintikka, J.: Knowledge and belief. Cornell University Press, Ithaca (1962)
4. Rantala, V.: Impossible worlds semantics and logical omniscience. Acta Philosophica Fennica 35, 106–115 (1982)

5. Fagin, R., Halpern, J.Y., Vardi, M.Y.: A non-standard approach to the logical omniscience problem. In: Parikh, R. (ed.) Theoretical Aspects of Reasoning about Knowledge: Proceedings of the Third Conference, pp. 41–55. Morgan Kaufmann, San Francisco (1990)
6. Fagin, R., Halpern, J.Y.: Belief, awareness and limited reasoning: Preliminary report. In: Proceedings of the 9th International Joint Conference on Artificial Intelligence, pp. 491–501 (1985)
7. Konolige, K.: A Deduction Model of Belief. Morgan Kaufmann, San Francisco (1986)
8. Elgot-Drapkin, J.J., Perlis, D.: Reasoning situated in time I: Basic concepts. Journal of Experimental and Theoretical Artificial Intelligence 2, 75–98 (1990)
9. Halpern, J.Y., Moses, Y., Vardi, M.Y.: Algorithmic knowledge. In: Fagin, R. (ed.) Theoretical Aspects of Reasoning about Knowledge: Proceedings of the Fifth Conference (TARK 1994), pp. 255–266. Morgan Kaufmann, San Francisco (1994)
10. Pucella, R.: Deductive algorithmic knowledge. J. Log. Comput. 16(2), 287–309 (2006)
11. Duc, H.N.: Logical omniscience vs. logical ignorance on a dilemma of epistemic logic. In: Pinto-Ferreira, C.A., Mamede, N.J. (eds.) EPIA 1995. LNCS, vol. 990, pp. 237–248. Springer, Heidelberg (1995)
12. Duc, H.N.: Reasoning about rational, but not logically omniscient, agents. Journal of Logic and Computation 7(5), 633–648 (1997)
13. Ågotnes, T., Alechina, N.: The dynamics of syntactic knowledge. Journal of Logic and Computation 17(1), 83–116 (2007)
14. Sierra, C., Godo, L., de Mántaras, R.L., Manzano, M.: Descriptive dynamic logic and its application to reflective architectu res. Future Gener. Comput. Syst. 12(2-3), 157–171 (1996)
15. Alechina, N., Jago, M., Logan, B.: Modal logics for communicating rule-based agents. In: Brewka, G., Coradeschi, S., Perini, A., Traverso, P. (eds.) Proceedings of the 17th European Conference on Artificial Intelligence (ECAI 2006), pp. 322–326. IOS Press, Amsterdam (2006)
16. Fisher, M., Ghidini, C.: Programming resource-bounded deliberative agents. In: Proceedings of the Sixteenth International Joint Conference on Artificial Intelligence (IJCAI 1999), pp. 200–205. Morgan-Kaufmann, San Francisco (1999)
17. Ågotnes, T.: A Logic of Finite Syntactic Epistemic States. Ph.D. thesis, Department of Informatics, University of Bergen, Norway (2004)
18. Ågotnes, T., Alechina, N.: Knowing minimum/maximum n formulae. In: Brewka, G., Coradeschi, S., Perini, A., Traverso, P. (eds.) Proceedings of the 17th European Conference on Artificial Intelligence (ECAI 2006), pp. 317–321. IOS Press, Amsterdam (2006)
19. Albore, A., Alechina, N., Bertoli, P., Ghidini, C., Logan, B., Serafini, L.: Model-checking memory requirements of resource-bounded reasoners. In: Proceedings of the Twenty-First National Conference on Artificial Intelligence (AAAI 2006), pp. 213–218. AAAI Press, Menlo Park (2006)
20. Alechina, N., Logan, B., Nguyen, H.N., Rakib, A.: Verifying time, memory and communication bounds in systems of reasoning agents. In: Padgham, L., Parkes, D., Müller, J., Parsons, S. (eds.) Proceedings of the Seventh International Conference on Autonomous Agents and Multiagent Systems (AAMAS 2008), Estoril, Portugal, May 2008, vol. 2, pp. 736–743. IFAAMAS (2008)
21. Ågotnes, T., Walther, D.: Towards a logic of strategic ability under bounded memory. In: Proceedings of the Workshop on Logics for Resource-Bounded Agents (2007)
22. Adjiman, P., Chatalic, P., Goasdoué, F., Rousset, M.C., Simon, L.: Scalability study of peer-to-peer consequence finding. In: Kaelbling, L.P., Saffiotti, A. (eds.) Proceedings of the Nineteenth International Joint Conference on Artificial Intelligence (IJCAI 2005), Edinburgh, Scotland, pp. 351–356. Professional Book Center (2005)
23. Emerson, E.A.: Temporal and modal logic. In: Handbook of Theoretical Computer Science. Formal Models and Sematics (B), vol. B, pp. 995–1072. Elsevier and MIT Press (1990)

24. Reynolds, M.: An axiomatization of full computation tree logic. Journal of Symbolic Logic 66(3), 1011–1057 (2001)
25. Alchourrón, C.E., Gärdenfors, P., Makinson, D.: On the logic of theory change: Partial meet functions for contraction and revision. Journal of Symbolic Logic 50, 510–530 (1985)
26. Baltag, A., Moss, L.S., Solecki, S.: The logic of public announcements, common knowledge, and private suspicions. In: Proceedings of the 7th conference on Theoretical aspects of rationality and knowledge, TARK 1998 (1998)

Normative Multi-agent Programs
and Their Logics

Mehdi Dastani[1], Davide Grossi[2], John-Jules Ch. Meyer[1], and Nick Tinnemeier[1]

[1]Universiteit Utrecht
The Netherlands
[2]Computer Science and Communication
University of Luxembourg, Luxembourg

Abstract. Multi-agent systems are viewed as consisting of individual agents whose behaviors are regulated by an organization artefact. This paper presents a simplified version of a programming language that is designed to implement norm-based artefacts. Such artefacts are specified in terms of norms being enforced by monitoring, regimenting and sanctioning mechanisms. The syntax and operational semantics of the programming language are introduced and discussed. A logic is presented that can be used to specify and verify properties of programs developed in this language.

1   Introduction

In this paper, multi-agent systems are considered as consisting of individual agents that are autonomous and heterogenous. Autonomy implies that each individual agent pursues its own objectives and heterogeneity implies that the internal states and operations of individual agents may not be known to external entities [19,7]. In order to achieve the overall objectives of such multi-agent systems, the observable/external behavior of individual agents and their interactions should be regulated/coordinated.

There are two main approaches to regulate the external behavior of individual agents. The first approach is based on coordination artefacts that are specified in terms of low-level coordination concepts such as synchronization of processes[16]. The second approach is motivated by organizational models, normative systems, and electronic institutions[17,13,7,10]. In such an approach, norm-based artefacts are used to regulate the behavior of individual agents in terms of norms being enforced by monitoring, regimenting and sanctioning mechanisms. Generally speaking, the social and normative perspective is conceived as a way to make the development and maintenance of multi-agent systems easier to manage. A plethora of social concepts (e.g., roles, social structures, organizations, institutions, norms) has been introduced in multi-agent system methodologies (e.g. Gaia [19]), models (e.g. OperA [6], $\mathcal{M}oise^+$ [11], electronic institutions and frameworks (e.g. AMELI [7], J-Moise+ [12]).

The main contribution of this paper is twofold. On the one hand, a simplified version of a programming language is presented that is designed to implement

J.-J.Ch. Meyer and J.M. Broersen (Eds.): KRAMAS 2008, LNAI 5605, pp. 16–31, 2009.

multi-agent systems in which the observable (external) behavior of individual agents is regulated by means of norm-based artefacts. Such artefacts are implemented in terms of social concepts such as norms and sanctions, monitor the actions performed by individual agents, evaluate their effects, and impose sanctions if necessary. On the other hand, we devise a logic to specify and verify properties of programs that implement norm-based artefacts.

In order to illustrate the idea of norm-based artefacts, consider the following simple example of a simulated train station where agents ought to buy a ticket before entering the platform or trains. To avoid the queue formation, agents are not checked individually before allowing them to enter the platform or trains. In this simulation, being on the platform without a ticket is considered as a violation and getting on the train without having a ticket is considered as a more severe violation. A norm-based artefact detects (all or some) violations by (all or some) agents and reacts on them by issuing a fine if the first violation occurs, for instance by charging the credit card of the defecting user, and a higher fine if the second violation occurs.

In this paper, we first briefly explain our idea of normative multi-agent systems and discuss two norm-based approaches to multi-agent systems, that is, ISLANDER/AMELI [7] and J-MOISE+ [12]. In section 3, we present the syntax and operational semantics of a programming language designed to implement normative multi-agent systems. This programming language allows the implementation of norm-based artefacts by providing programming constructs to represent norms and mechanisms to enforce them. In section 4, a logic is presented that can be used to specify and verify properties of norm-based artefacts implemented in the presented programming language. Finally, in section 5, we conclude the paper and discuss some future directions in this research area.

2   Norms and Multi-Agent Systems

Norms in multi-agent systems can be used to specify the standards of behavior that agents ought to follow to meet the overall objectives of the system. However, to develop a multi-agent system does not boil down to state a number of standards of behavior in the form of a set of norms, but rather to organize the system in such a way that those standards of behavior are actually followed by the agents. This can be achieved by regimentation [13] or enforcement mechanisms, e.g., [10].

When regimenting norms all agents' external actions leading to a violation of those norms are made impossible. Via regimentation (e.g., gates in train stations) the system prevents an agent from performing a forbidden action (e.g., entering a train platform without a ticket). However, regimentation drastically decreases agent autonomy. Instead, enforcement is based on the idea of responding after a violation of the norms has occurred. Such a response, which includes sanctions, aims to return the system to an acceptable/optimal state. Crucial for enforcement is that the actions that violate norms are observable by the system (e.g., fines can be issued only if the system can detect travelers entering the platform or trains without a ticket). Another advantage of having enforcement over regimentation is that allowing for violations contributes to the flexibility and autonomy of the agent's behavior [3]. These norms are often specified by means of concepts like permissions, obligations, and prohibitions.

In the literature of multi-agent systems related work can be found on electronic institutions. In particular, ISLANDER[8] is a formal framework for specifying norms in institutions, which is used in the AMELI platform [7] for executing electronic institutions based on norms provided in it. However, the key aspect of ISLANDER/AMELI is that norms can never be violated by agents. In other words, systems programmed via ISLANDER/AMELI make only use of regimentation in order to guarantee the norms to be actually followed. Similar to some extensions of electronic institutions (cf. [9,18]) we relax this assumption, guaranteeing higher autonomy to the agents, and higher flexibility to the system. In contrast to our work, however, the norms of [7,9,18] all relate to (dialogical) actions the agents should or should not perform and ignore the issue of expressing more high-level norms concerning a state of the environment that should be brought about. Such high-level norms can be used to represent *what* the agents should establish – in terms of a declarative description of a state – rather than specifying *how* they should establish it.

A work that is also concerned with programming multiagent systems using (among others) normative concepts is J-MOISE+ [12], which is an organizational middleware that follows the Moise+ model[11]. This approach, like ours, builds on programming constructs investigated in social and organizational sciences. In contrast to [7,9,18], J-MOISE+ is concerned with more high-level norms that are about declarative descriptions of a state that should be achieved. However, norms in S-MOISE+ typically lack monitoring and sanctioning mechanisms for their implementation which are the focus of our proposal. It should be noted that [14] advocates the use of artifacts to implement norm enforcement mechanisms. However, it is not explained how such a mechanism is to be realized.

To summarise, in the work on electronic institutions norms pertain to very low-level procedures that directly refer to actions, whereas the work on J-MOISE+ is concerned with more high-level norms pertaining to declarative descriptions of the system, but no specific system's response to violations is built in the framework. Another important issue is that the above mentioned approaches lack a formal operational semantics [15], which we deem very important for a thorough understanding and analysis of the proposed concepts. The present paper fills this gap along the same lines that have been followed for the operationalization of BDI notions in the APL-like agent programming languages [4,5]. Finally, it should be noted that besides normative concepts MOISE+ and ISLANDER/AMELI also provide a variety of other social and organizational concepts. Since the focus of this paper is on the normative aspect, the above discussion is limited hereto. Future research will focus on other social and organizational concepts.

3   Programming Multi-Agent Systems with Norms

In this section, we present a programming language to facilitate the implementation of multi-agent systems with norms, i.e., to facilitate the implementation of norm-based artefacts that coordinate/regulate the behavior of participating individual agents. A normative multi-agent system (i.e., a norm-based artefact) is considered to contain two modules: an organization module that specifies norms and sanctions, and an environment module in which individual agents can perform actions. The individual agents are assumed to be implemented in a programming language, not necessarily known to the multi-agent system programmer, though the programmer is required to have the reference to the (executable) programs of each individual agent. It is also assumed that all actions that are performed by individual agents are observable to the multi-agent system (i.e., norm-based artefact). Note that the reference to the (executable) programs of individual agents are required such that multi-agent systems (i.e., normative artefact) can observe the actions generated by the agent programs. Finally, we assume that the effect of an individual agent's action in the external environment is determined by the program that implements the norm-based artefact (i.e., by the multi-agent system program). Most noticeably it is not assumed that the agents are able to reason about the norms of the system.

The programming language for normative multi-agent systems provides programming constructs to specify the effect of an agent's actions in the environment, norms, sanctions, and the initial state of the environment. Moreover, the programming language is based on a monitoring and a sanctioning mechanism that observes the actions performed by the agents, determines their effects in the shared environment, determines the violations caused by performing the actions, and possibly, imposes sanctions. A program in this language is the implementation of a norm-based artefact. As we assume that the norm-based artefacts determine the effects of external actions in the shared environment, the programming language should provide constructs to implement these effects. The effect of an agent's (external) actions is specified by a set of literals that should hold in the shared environment after the external action is performed by the agent. As external actions can have different effects when they are executed in different states of the shared environment, we add a set of literals that function as the pre-condition of those effect.

We consider norms as being represented by counts-as rules [17], which ascribe "institutional facts" (e.g. "a violation has occurred"), to "brute facts" (e.g. "an agent is on the train without ticket"). For example, a counts-as rule may express the norm "an agent on the train without ticket counts-as a violation". In our framework, brute facts constitute the environment shared by the agents, while institutional facts constitute the normative/institutional state of the multi-agent system. Institutional facts are used with the explicit aim of triggering system's reactions (e.g., sanctions). As showed in [10] counts-as rules can enjoy a rather classical logical behavior, and are here implemented as simple rules that relate brute and normative facts. In the presented programming language, we distinguish brute facts from normative (institutional) facts and assume two disjoint sets of propositions to denote these facts.

Brute and institutional facts constitute the (initial) state of the multi-agent system (i.e., the state of the norm-based artefact). Brute facts are initially set by the programmer by means of the initial state of the shared environment. These facts can change as individual agents perform actions in the shared environment. Normative facts are determined by applying counts-as rules in multi-agent states. The application of counts-as rules in subsequent states of a multi-agent system realizes a monitoring mechanism as it determines and detects norm violations during the execution of the multi-agent system.

Sanctions are also implemented as rules, but follow the opposite direction of counts-as rules. A sanction rule determines which brute facts will be brought about by the system as a consequence of the normative facts. Typically, such brute facts are sanctions, such as fines. Notice that in human systems sanctions are usually issued by specific agents (e.g. police agents). This is not the case in our computational setting, where sanctions necessarily follow the occurrence of a violation if the relevant sanction rule is in place (comparable to automatic traffic control and issuing tickets). It is important to stress, however, that this is not an intrinsic limitation of our approach. We do not aim at mimicking human institutions but rather providing the specification of computational systems.

3.1 Syntax

In order to represent brute and institutional facts in our normative multi-agent systems programming language, we introduce two disjoint sets of propositions to denote these facts. The syntax of the normative multi-agent system programming language is presented in Figure 1 using the EBNF notation. In the following, we use `<b-prop>` and `<i-prop>` to be propositional formulae taken from two different disjoint sets of propositions. Moreover, we use `<ident>` to denote a string and `<int>` to denote an integer.

In order to illustrate the use of this programming language, consider the following underground station example.

```
Agents:           passenger   PassProg   1
Facts:            {-at_platform, -in_train, -ticket}
Effects:          {-at_platform} enter {at_platform},
                  {-ticket} buy_ticket {ticket},
                  {at_platform, -in_train} embark
                               {-at_platform, in_train}
Counts_as rules: {at_platform , -ticket} ⇒ {viol₁},
                  {in_train , -ticket} ⇒ {viol⊥}
Sanction rules:  {viol₁} ⇒ {fined₁₀}
```

This program creates one agent called **passenger** whose (executable) specification is included in a file with the name **PassProg**. The **Facts**, which implement

```
N-MAS_Prog    := "Agents: " (<agentName> <agentProg> [<nr>])+ ;
                 "Facts: " <bruteFacts>
                 "Effects: " <effects>
                 "Counts-as rules: " <counts-as>
                 "Sanction rules: " <sanctions>;
<agentName>   := <ident>;
<agentProg>   := <ident>;
<nr>          := <int>;
<bruteFacts>  := <b-literals>;
<effects>     := ({<b-literals>} <actionName> {<b-literals>})+;
<counts-as>   := ( <literals> ⇒ <i-literals> )+;
<sanctions>   := ( <i-literals> ⇒ <b-literals>)+;
<actionName>  := <ident>;
<b-literals>  := <b-literal> {"," <b-literal>};
<i-literals>  := <i-literal> {"," <i-literal>};
<literals>    := <literal> {"," <literal>};
<literal>     := <b-literal> | <i-literal>;
<b-literal>   := <b-prop> | "not" <b-prop>;
<i-literal>   := <i-prop> | "not" <i-prop>;
```

Fig. 1. The EBNF syntax of the normative multi-agent programming language

brute facts, determine the initial state of the shared environment. In this case, the agent is not at the platform (-at_platform) nor in the train (-in_train) and has no ticket (-ticket). The Effects indicate how the environment can advance in its computation. Each effect is of the form {pre-condition} action {post-condition}. The first effect, for instance, means that if the agent performs an enter action when not at the platform, the result is that the agent is on the platform (either with or without a ticket). Only those effects that are changed are thus listed in the post-condition. The Counts_as rules determine the normative effects for a given (brute and normative) state of the multi-agent system. The first rule, for example, states that being on the platform without having a ticket is a specific violation (marked by $viol_1$). The second rule marks states where agents are on a train without a ticket with the specifically designated literal $viol_\perp$. This literal is used to implement regimentation. The operational semantics of the language ensures that the designated literal $viol_\perp$ can never hold during any run of the system (see Definition 3). Intuitively, rules with $viol_\perp$ as consequence could be thought of as placing gates blocking an agent's action. Finally, the aim of Sanction rules is to determine the punishments that are imposed as a consequence of violations. In the example the violation of type $viol_1$ causes the sanction $fined_{10}$ (e.g., a 10 EUR fine).

Counts-as rules obey syntactic constraints. Let $l = (\Phi \Rightarrow \Psi)$ be a rule, we use $cond_l$ and $cons_l$ to indicate the condition Φ and consequent Ψ of the rule l, respectively. We consider only sets of rules such that 1) they are finite; 2) they are such that each condition has exactly one associated consequence (i.e., all the consequences of a given conditions are packed in one single set cons); and 3) they are such that for counts-as rule k, l, if $cons_k \cup cons_l$ is inconsistent (i.e., contains p and $-p$), then $\mathsf{cond}_k \cup \mathsf{cond}_l$ is also inconsistent. That is to say, rules trigger inconsistent conclusions only in different states. In the rest of this paper, sets of rules enjoying these three properties are denoted by **R**.

Note that in the current approach a multi-agent system consists of only one normative artifact (environment) the agents interact with, which suggests a centralized approach. However, extending the notion of a multi-agent system to contain a multitude of normative artifacts the agents interact with does not affect the norm enforcement mechanism of the individual artifacts. Because this paper concentrates on the norm enforcement mechanism of a normative artifact we decided to adopt the assumption of one environment in order not to complicate matters further. Our framework can be extended with a set of normative artifacts, each of which is responsible for monitoring a specific set of actions and sanctioning the corresponding agents.

3.2   Operational Semantics

One way to define the semantics of this programming language is by means of operational semantics. Using such semantics, one needs to define the configuration (i.e., state) of normative multi-agent systems and the transitions that such configurations can undergo through transition rules. The state of a multi-agent system with norms consists of the state of the external environment, the normative state, and the states of individual agents.

Definition 1. *(Normative Multi-Agent System Configuration) Let P_b and P_n be two disjoint sets of literals denoting atomic brute and normative facts (including* $\mathsf{viol}_\perp$*), respectively. Let A_i be the configuration of individual agent i. The configuration of a normative multi-agent system is defined as $\langle \mathcal{A}, \sigma_b, \sigma_n \rangle$ where $\mathcal{A} = \{A_1, \ldots, A_n\}$, σ_b is a consistent set of literals from P_b denoting the brute state of multi-agent system and σ_n is a consistent set of literals from P_n denoting the normative state of multi-agent system.*

The configuration of such a multi-agent system can change for various reasons, e.g., because individual agents perform actions in the external environment or because the external environment can have its own internal dynamics (the state of a clock changes independent of an individual agent's action). In operational semantics, transition rules specify how and when configurations can change, i.e., they specify which transition between configurations are allowed and when they can be derived. In this paper, we consider only the transition rules that specify the transition of multi-agent system configurations as a result of performing external actions by individual agents. Of course, individual agents can perform (internal) actions that modify only their own configurations and have no influence on the multi-agent system configuration. The transition rules to derive such transitions are out of the scope of this paper.

Definition 2. *(Transitions of Individual Agent's Actions) Let A_i and A'_i be configurations of individual agent i, and $\alpha(i)$ be an (observable) external action performed by agent i. Then, the following transition captures the execution of an external action by an agent.*

$$A_i \xrightarrow{\alpha(i)} A_i' \ : \ agent \ i \ can \ perform \ external \ action \ \alpha$$

This transition indicates that an agent configuration can change by performing an external action. The performance of the external action is broadcasted to the multi-agent system level. Note that no assumption is made about the internals of individual agents as we do not present transition rules for deriving internal agent transitions (denoted as $A \longrightarrow A'$). The only assumption is that the action of the agent is observable. This is done by labeling the transition with the external action name.

Before presenting the transition rule specifying the possible transitions of the normative MAS configurations, the closure of a set of conditions under a set of (counts-as and sanction) rules needs to be defined. Given a set $\mathbf{R}$ of rules and a set X of literals, we define the set of applicable rules in X as $\mathtt{Appl}^{\mathbf{R}}(X) = \{\Phi \Rightarrow \Psi \mid X \models \Phi\}$. The closure of X under $\mathbf{R}$, denoted as $\mathtt{Cl}^{\mathbf{R}}(X)$, is inductively defined as follows:

B: $\mathtt{Cl}_0^{\mathbf{R}}(X) = X \cup \left(\bigcup_{l \in \mathtt{Appl}^{\mathbf{R}}(X)} \mathtt{cons}_l \right)$

S: $\mathtt{Cl}_{n+1}^{\mathbf{R}}(X) = \mathtt{Cl}_n^{\mathbf{R}}(X) \cup \left(\bigcup_{l \in \mathtt{Appl}^{\mathbf{R}}(\mathtt{Cl}_n^{\mathbf{R}}(X))} \mathtt{cons}_l \right)$

Because of the properties of finiteness, consequence uniqueness and consistency of $\mathbf{R}$ one and only one finite number $m + 1$ can always be found such that $\mathtt{Cl}_{m+1}^{\mathbf{R}}(X) = \mathtt{Cl}_m^{\mathbf{R}}(X)$ and $\mathtt{Cl}_m^{\mathbf{R}}(X) \neq \mathtt{Cl}_{m-1}^{\mathbf{R}}(X)$. Let such $m + 1$ define the closure X under $\mathbf{R}$: $\mathtt{Cl}^{\mathbf{R}}(X) = \mathtt{Cl}_{m+1}^{\mathbf{R}}(X)$. Note that the closure may become inconsistent due to the ill-defined set of counts-as rules. For example, the counts-as rule $p \Rightarrow -p$ (or the set of counts as rules $\{p \Rightarrow q \ , \ q \Rightarrow -p\}$), where p and q are normative facts, may cause the normative state of a multi-agent system to become inconsistent.

We can now define a transition rule to derive transitions between normative multi-agent system configurations. In this transition rule, the function up determines the effect of action $\alpha(i)$ on the environment σ_b based on its specification $(\Phi \ \alpha(i) \ \Phi')$ as follows:

$$up(\alpha(i), \sigma_b) = (\sigma_b \cup \Phi') \setminus (\{p \mid -p \in \Phi'\} \cup \{-p \mid p \in \Phi'\})$$

Definition 3. *(Transition Rule for Normative Multi-Agent Systems) Let $\mathbf{R_c}$ be the set of counts-as rules, $\mathbf{R_s}$ be the set of sanction rules, and $(\Phi \ \alpha(i) \ \Phi')$ be the specification of action $\alpha(i)$. The multi-agent transition rule for the derivation of normative multi-agent system transitions is defined as follows:*

$$\frac{A_i \in \mathcal{A} \ \ \& \ \ A_i \xrightarrow{\alpha(i)} A_i' \ \ \& \ \ \sigma_b \models \Phi \ \ \& \ \ \sigma_b' = up(\alpha(i), \sigma_b)}{\langle \mathcal{A}, \sigma_b, \sigma_n \rangle \longrightarrow \langle \mathcal{A}', \sigma_b' \cup S, \sigma_n' \rangle}$$

$$\sigma_n' = \mathtt{Cl}^{\mathbf{R_c}}(\sigma_b') \setminus \sigma_b' \ \ \& \ \ \sigma_n' \not\models \mathrm{viol}_\perp \ \ \& \ \ S = \mathtt{Cl}^{\mathbf{R_s}}(\sigma_n') \setminus \sigma_n' \ \ \& \ \ \sigma_b' \cup S \not\models \perp$$

where $\mathcal{A}' = (\mathcal{A} \setminus \{A_i\}) \cup \{A_i'\}$ and $\mathrm{viol}_\perp$ *is the designated literal for regimentation.*

This transition rule captures the effects of performing an external action by an individual agent on both external environments and the normative state of the

MAS. First, the effect of α on σ_b is computed. Then, the updated environment is used to determine the new normative state of the system by applying all counts-as rules to the new state of the external environments. Finally, possible sanctions are added to the new environment state by applying sanction rules to the new normative state of the system. In should be emphasized that other multi-agent transition rules, such as transition rules for communication actions, are not presented in this paper because the focus here is on how norms determine the effects of external actions.

Note that the external action of an agent can be executed only if it would not result in a state containing $viol_\perp$. This captures exactly the regimentation of norms. Hence, once assumed that the initial normative state does not include $viol_\perp$, it is easy to see that the system will never be in a $viol_\perp$-state. It is important to note that when a normative state σ'_n becomes inconsistent, the proposed transition rule cannot be applied because an inconsistent σ'_n entails $viol_\perp$. Also, note that the condition $\sigma'_b \cup S \not\models \perp$ guarantees that the environment state never can become inconsistent. Finally, it should be emphasized that the normative state σ'_b is not defined on σ_n and is always computed anew.

4   Logic

In this section, we propose a logic to specify and verify liveness and safety properties of multi-agent system programs with norms. This logic, which is a variant of Propositional Dynamic Logic (PDL, see [2]), is in the spirit of [1] and rely on that work. It is important to note that the logic developed in [1] aims at specifying and verifying properties of single agents programmed in terms of beliefs, goals, and plans. Here we modify the logic and apply it to multi-agent system programs. We first introduce some preliminaries before presenting the logic.

4.1   Preliminaries

We show how the programming constructs can be used for grounding a logical semantics. Let P denote the set of propositional variables used to describe brute and normative states of the system. It is assumed that each propositional variable in P denotes either an institutional/normative or a brute state-of-affairs: $P = P_n \cup P_b$ and $P_n \cap P_b = \emptyset$. A state s is represented as a pair $\langle \sigma_b, \sigma_n \rangle$ where $\sigma_b = \{(-)p_1, \ldots, (-)p_n : p_i \in P_b\}$ is a consistent set of literals (i.e., for no $p \in P_b$ it is the case that $p \in \sigma_b$ and $-p \in \sigma_b$), and σ_n is like σ_b for P_n.

Rules are pairs of conditions and consequences $(\{(-)p_1, \ldots, (-)p_n \mid (-)p_i \in X\}, \{(-)q_1, \ldots, (-)q_k \mid (-)q_i \in Y\})$ with X and Y being either σ_b or σ_n when applied in state $\langle \sigma_b, \sigma_n \rangle$. Following [10], if $X = \sigma_b$ and $Y = \sigma_n$ then the rule is called *bridge counts-as rule*; if $X = Y = \sigma_n$ then the rule is an *institutional counts-as rule*; if $X = \sigma_n$ and $Y = \sigma_b$ then the rule is a *sanction rule*. Literals p's and q's are taken to be disjoint. Leaving technicalities aside, bridge counts-as rules connect brute states to normative/institutional ones, institutional counts-as rules connect institutional facts to institutional facts, and sanction rules connect normative states to brute ones.

Given a set $\mathbf{R}$ of rules, we say a state $s = \langle \sigma_b, \sigma_n \rangle$ to be $\mathbf{R}$-aligned if for all pairs $(\mathtt{cond}_k, \mathtt{cons}_k)$ in $\mathbf{R}$: if $\mathtt{cond}_k$ is satisfied by $\sigma_b \cup \sigma_n$ (σ_b in the case of a bridge counts-as rule and σ_n in the case of an institutional counts-as or a sanction rule), then $\mathtt{cons}_k$ is satisfied by σ_n (in the case of a bridge or institutional counts-as rule) or by σ_b (in the case of a sanction rule), respectively. States that are $\mathbf{R}$-aligned are states which instantiate the normative system specified by $\mathbf{R}$.

Let the set of agents' external actions $\mathtt{Ac}$ be the union $\bigcup_{i \in I} \mathtt{Ac}_i$ of the finite sets $\mathtt{Ac}_i$ of external actions of each agent i in the set I. We denote external actions as $\alpha(i)$ where $\alpha \in \mathtt{Ac}_i$ and $i \in I$. We associate now with each $\alpha(i) \in \mathtt{Ac}_i$ a set of pre- and post-conditions $\{(-)p_1 \in \sigma_b, \ldots, (-)p_n \in \sigma_b\}$, $\{(-)q_1 \in \sigma'_b, \ldots, (-)q_k \in \sigma'_b\}$ (where p's and q's are not necessarily disjoint) when $\alpha(i)$ is executed in a state with brute facts set σ_b which satisfies the pre-condition then the resulting state s' has the brute facts set σ'_b which satisfies the post-condition (including replacing p with $-p$ if necessary to preserve consistency) and it is such that *the rest of σ'_b is the same as σ_b.* Executing an action $\alpha(i)$ in different configurations may give different results. For each $\alpha(i)$, we denote the set of pre- and post-condition pairs $\{(\mathtt{prec}_1, \mathtt{post}_1), \ldots, (\mathtt{prec}_m, \mathtt{post}_m)\}$ by $C_b(\alpha(i))$. We assume that $C_b(\alpha(i))$ is finite, that pre-conditions $\mathtt{prec}_k, \mathtt{prec}_l$ are mutually exclusive if $k \neq l$, and that each pre-condition has exactly one associated post-condition. We denote the set of all such pre- and post-conditions of all agents' external actions by $\mathbf{C}$.

Now everything is put into place to show how the execution of $\alpha(i)$ in a state with brute facts set σ_b also univocally changes the normative facts set σ_n by means of the applicable counts-as rules, and adds the resulting sanctions by means of the applicable sanction rules. If $\alpha(i)$ is executed in a state $\langle \sigma_b, \sigma_n \rangle$ with brute facts set σ_b, which satisfies the pre-conditions, then the resulting state $\langle \sigma'_b \cup S, \sigma'_n \rangle$ is such that σ'_b satisfies the brute post-condition of $\alpha(i)$ (including replacing p with $-p$ if necessary) and the rest of σ'_b is the same of σ_b; σ'_n is determined by the closure of σ'_b with counts-as rules $\mathbf{R}_c$; sanctions S are obtained via closure of σ'_n with sanction rules $\mathbf{R}_s$.

4.2   Language

The language L for talking about normative multi-agent system programs is just the language of PDL built out of a finite set of propositional variables $P \cup -P$ (i.e., the literals built from P), used to describe the system's normative and brute states, and a finite set $\mathtt{Ac}$ of agents' actions. Program expressions ρ are built out of external actions $\alpha(i)$ as usual, and formulae ϕ of L are closed under boolean connectives and modal operators:

$$\rho ::= \alpha(i) \mid \rho_1 \cup \rho_2 \mid \rho_1; \rho_2 \mid ?\phi \mid \rho^*$$
$$\phi ::= (-)p \mid \neg\phi \mid \phi_1 \wedge \phi_2 \mid \langle \rho \rangle \phi$$

with $\alpha(i) \in \mathtt{Ac}$ and $(-)p \in P \cup -P$. Connectives $\vee$ and $\rightarrow$, and the modal operator $[\rho]$ are defined as usual.

4.3   Semantics

The language introduced above is interpreted on transition systems that generalize the operational semantics presented in the earlier section, in that they do not describe a particular program, but all possible programs —according to **C**— generating transitions between all the $\mathbf{R_c}$ and $\mathbf{R_s}$-aligned states of the system. As a consequence, the class of transition systems we are about to define will need to be parameterized by the sets $\mathbf{C}$, $\mathbf{R_c}$ and $\mathbf{R_s}$.

A model is a structure $M = \langle S, \{R_{\alpha(i)}\}_{\alpha(i) \in \mathtt{Ac}}, V \rangle$ where:

- S is a set of $\mathbf{R_c}$ and $\mathbf{R_s}$-aligned states.
- $V = (V_b, V_n)$ is the evaluation function consisting of brute and normative valuation functions V_b and V_n such that for $s = \langle \sigma_b, \sigma_n \rangle$, $V_b(s) = \sigma_b$ and $V_n(s) = \sigma_n$.
- $R_{\alpha(i)}$, for each $\alpha(i) \in \mathtt{Ac}$, is a relation on S such that $(s, s') \in R_{\alpha(i)}$ iff for some $(\mathtt{prec}_k, \mathtt{post}_k) \in C(\alpha(i))$, $\mathtt{prec}_k(s)$ and $\mathtt{post}_k(s')$, i.e., for some pair of pre- and post-conditions of $\alpha(i)$, the pre-condition holds for s and the corresponding post-condition holds for s'. Note that this implies two things. First, an $\alpha(i)$ transition can only originate in a state s which satisfies one of the pre-conditions for $\alpha(i)$. Second, since pre-conditions are mutually exclusive, every such s satisfies exactly one pre-condition, and all $\alpha(i)$-successors of s satisfy the matching post-condition.

Given the relations corresponding to agents' external actions in M, we can define sets of paths in the model corresponding to any PDL program expression ρ in M. A set of paths $\tau(\rho) \subseteq (S \times S)^*$ is defined inductively:

- $\tau(\alpha(i)) = \{(s, s') : R_{\alpha(i)}(s, s')\}$
- $\tau(\phi?) = \{(s, s) : M, s \models \phi\}$
- $\tau(\rho_1 \cup \rho_2) = \{z : z \in \tau(\rho_1) \cup \tau(\rho_2)\}$
- $\tau(\rho_1; \rho_2) = \{z_1 \circ z_2 : z_1 \in \tau(\rho_1), \; z_2 \in \tau(\rho_2)\}$, where $\circ$ is concatenation of paths , such that $z_1 \circ z_2$ is only defined if z_1 ends in the state where z_2 starts
- $\tau(\rho^*)$ is the set of all paths consisting of zero or finitely many concatenations of paths in $\tau(\rho)$ (same condition on concatenation as above)

Constructs such as If ϕ then ρ_1 else ρ_2 and while ϕ do ρ are defined as $(\phi?; \rho_1) \cup (\neg\phi?; \rho_2)$ and $(\phi?; \rho)^*; \neg\phi$, respectively. The satisfaction relation $\models$ is inductively defined as follows:

- $M, s \models (-)p$ iff $(-)p \in V_b(s)$ for $p \in P_b$
- $M, s \models (-)p$ iff $(-)p \in V_n(s)$ for $p \in P_n$
- $M, s \models \neg\phi$ iff $M, s \not\models \phi$
- $M, s \models \phi \wedge \psi$ iff $M, s \models \phi$ and $M, s \models \psi$
- $M, s \models \langle \rho \rangle \phi$ iff there is a path in $\tau(\rho)$ starting in s which ends in a state s' such that $M, s' \models \phi$.
- $M, s \models [\rho]\phi$ iff for all paths $\tau(\rho)$ starting in s, the end state s' of the path satisfies $M, s' \models \phi$.

Let the class of transition systems defined above be denoted $\mathbf{M_{C,R_c,R_s}}$ where $\mathbf{C}$ is the set of pre- and post-conditions of external actions, $\mathbf{R_c}$ is the set of counts-as rules and $\mathbf{R_s}$ the set of sanction rules.

4.4   Axiomatics

The axiomatics shows in what the logic presented differs w.r.t. standard PDL. In fact, it is a conservative extension of PDL with domain-specific axioms needed to axiomatize the behavior of normative multi-agent system programs.

For every pre- and post-condition pair $(\mathbf{prec}_i, \mathbf{post}_i)$ we describe states satisfying $\mathbf{prec}_i$ and states satisfying $\mathbf{post}_i$ by formulas of L. More formally, we define a formula $tr(X)$ corresponding to a pre- or post-condition X as follows: $tr((-)p) = (-)p$ and $tr(\{\phi_1, \ldots, \phi_n\}) = tr(\phi_1) \wedge \ldots \wedge tr(\phi_n)$. This allows us to axiomatize pre- and post-conditions of actions. The conditions and consequences of counts-as rules and sanction rules can be defined in similar way as pre- and post-conditions of actions, respectively. The set of models $\mathbf{M}_{\mathbf{C}, \mathbf{R}_c, \mathbf{R}_s}$ is axiomatized as follows:

PDL. Axioms and rules of PDL
Ax Consistency. Consistency of literals: $\neg(p \wedge -p)$
Ax Counts-as. For every rule $(\mathbf{cond}_k, \mathbf{cons}_k)$ in $\mathbf{R}_c$: $tr(\mathbf{cond}_k) \rightarrow tr(\mathbf{cons}_k)$
Ax Sanction. For every rule $(\mathbf{viol}_k, \mathbf{sanc}_k)$ in $\mathbf{R}_s$: $tr(\mathbf{viol}_k) \rightarrow tr(\mathbf{sanc}_k)$
Ax Regiment. $\mathbf{viol}_\perp \rightarrow \perp$
Ax Frame. For every action $\alpha(i)$ and every pair of pre- and post-conditions $(\mathbf{prec}_j, \mathbf{post}_j)$ in $C(\alpha(i))$ and formula Φ built out of P_b not containing any propositional variables occurring in $\mathbf{post}_j$:
$$tr(\mathbf{prec}_j) \wedge \Phi \rightarrow [\alpha(i)](tr(\mathbf{post}_j) \wedge \Phi)$$
This is a frame axiom for actions.
Ax Non-Executability. For every action $\alpha(i)$, where all possible pre-conditions in $C(\alpha(i))$ are $\mathbf{prec}_1, \ldots, \mathbf{prec}_k$: $\neg tr(\mathbf{prec}_1) \wedge \ldots \wedge \neg tr(\mathbf{prec}_k) \rightarrow \neg \langle \alpha(i) \rangle \top$
where $\top$ is a tautology.
Ax Executability. For every action $\alpha(i)$ and every pre-condition $\mathbf{prec}_j$ in $C(\alpha(i))$: $tr(\mathbf{prec}_j) \rightarrow \langle \alpha(i) \rangle \top$

Let us call the axiom system above $\mathbf{Ax}_{\mathbf{C}, \mathbf{R}_c, \mathbf{R}_s}$, where $\mathbf{C}$ is the set of brute pre- and post-conditions of atomic actions, $\mathbf{R}_c$ is the set of counts-as rules, and $\mathbf{R}_s$ is the set of sanction rules.

Theorem 1. *Axiomatics* $\mathbf{Ax}_{\mathbf{C}, \mathbf{R}_c, \mathbf{R}_s}$ *is sound and weakly complete for the class of models* $\mathbf{M}_{\mathbf{C}, \mathbf{R}_c, \mathbf{R}_s}$.

Proof. Soundness is proven as usual by induction on the length of derivations. We sketch the proof of completeness. It builds on the usual completeness proof of PDL via finite canonical models. Given a consistent formula ϕ to be proven satisfiable, such models are obtained via the Fischer-Ladner closure of the set of subformulae of the formula ϕ extended with all pre- and post-conditions of any action $\alpha(i)$ occurring in ϕ. Let $FLC(\phi)$ denote such closure. The canonical model consists of all the maximal $\mathbf{Ax}_{\mathbf{C}, \mathbf{R}_c, \mathbf{R}_s}$-consistent subsets of $FLC(\phi)$. The accessibility relation and the valuation of the canonical model are defined like in PDL and the truth lemma follows in the standard way. It remains to be proven that the model satisfies the axioms. First, since the states in the model are maximal and consistent w.r.t. *Ax Counts-as*, *Ax Sanction*, *Ax Consistency*, and *AxRegiment*, they are $\mathbf{R_c}$- and $\mathbf{R_s}$-aligned, σ_b and σ_n are consistent, and no state is such that $\sigma_n \models \mathtt{viol}_\perp$. Second, it should be shown that the canonical model satisfies the pre- and post-conditions of the actions occurring in ϕ in that: a) no action $\alpha(i)$ is executable in a state s if none of its preconditions are satisfied by s, and b) if they hold in s then the corresponding post-conditions hold in s' which is accessible by $R_{\alpha(i)}$ from s. As to a), if a state s in the canonical model does not satisfy any of the preconditions of $\alpha(i)$ then, by *Ax Non-Executability* and the definition of the canonical accessibility relation, there is no s' in the model such that $sR_{\alpha(i)}s'$. As to b), if a state s in the canonical model satisfies one of the preconditions $\mathtt{prec}_j$ of $\alpha(i)$ then $tr(\mathtt{prec}_j)$ belongs to s and, by *Ax Frame*, $[\alpha(i)]\,tr(\mathtt{post}_j)$ also do. Now, *Ax Executability* guarantees that there exists at least one s' such that $sR_{\alpha(i)}s'$, and, for any s' such that $sR_{\alpha(i)}s'$, by the definition of such canonical accessibility relation, s' contains $tr(\mathtt{post}_j)$ (otherwise it would not be the case that $sR_{\alpha(i)}s'$). On the other hand, for any literal $(-)p$ in s not occurring in $tr(\mathtt{post}_j)$, its value cannot change from s to s' since, if it would, then for *Ax Frame* it would not be the case that $sR_{\alpha(i)}s'$, which is impossible. This concludes the proof.

4.5   Verification

To verify a normative multi-agent system program means, in our perspective, to check whether the program implementing the normative artefact is soundly designed w.r.t. the regimentation and sanctioning mechanisms it is supposed to realize or, to put it in more general terms, to check whether certain property holds in all (or some) states reachable by the execution traces of the multi-agent system program. In order to do this, we need to translate a multi-agent system program into a PDL program expression.

As explained in earlier sections, a multi-agent system program assumes a set of behaviors $A_1, \ldots, A_n$ of agents $1, \ldots, n$, each of which is a sequence of external actions (the agents actions observed from the multi-agent level), i.e., $A_i = \alpha_i^1; \alpha_i^2, \ldots$ where $\alpha_i^j \in Ac$. [1] Moreover, a multi-agent system program with norms consists of an initial set of brute facts, a set of counts-as rules and a set of sanction rules which together determine the initial state of the program. In this paper, we consider the execution of a multi-agent program as interleaved executions of the involved agents' behaviors started at the initial state.

Given I as the set of agents' names and A_i as the behavior of agent $i \in I$, the execution of a multi-agent program can be described as PDL expression $\bigcup interleaved(\{A_i | i \in I\})$, where $interleaved(\{A_i | i \in I\})$ yields all possible interleavings of agents' behaviors, i.e., all possible interleavings of actions from sequences A_i. It is important to notice that $\bigcup interleaved(\{A_i | i \in I\})$ corresponds to the set of computations sequences (execution traces) generated by the operational semantics.

---

[1] Note an agent's behavior can always be written as a (set of) sequence(s) of actions, which in turn can be written as a PDL expressions.

The general verification problem can now be formulated as follows. Given a multi-agent system program with norms in a given initial state satisfying $\phi \in L$, the state reached after the execution of the program satisfies ψ, i.e.:

$$\phi \rightarrow \langle [\bigcup interleaved(\{A_i | i \in I\})] \rangle \psi$$

In the above formulation, the modality $\langle [\ldots] \rangle$ is used to present both safety [...] and liveness $\langle \ldots \rangle$ properties. We briefly sketch a sample of such properties using again the multi-agent system program with norms which implements the train station example with one passenger agent (see Section 3).

Sanction follows violation. Entering without a ticket results in a fine, i.e.,

$$-\text{at\_platform} \wedge -\text{train} \wedge -\text{ticket} \rightarrow [\text{enter}](\text{viol}_1 \wedge \text{pay}_{10}).$$

Norm obedience avoids sanction. Buying a ticket if you have none and entering the platform does not result in a fine, i.e.:

$$-\text{at\_platform} \wedge -\text{train} \rightarrow \langle \text{ If} -\text{ticket then } \text{buy\_ticket; enter} \rangle \, (\text{at\_platform} \wedge -\text{pay}_{10}).$$

Regimentation. It is not possible for an agent to enter the platform and embark the train without a ticket, i.e.:

$$-\text{at\_platform} \wedge -\text{train} \wedge -\text{ticket} \rightarrow [\text{enter; embark}] \bot$$

Note that there is only one passenger agent involved in the example program. For this property, we assume that the passenger's behavior is enter; embark. Note also that:

$$\bigcup interleaved(\{\text{enter; embark}\}) = \text{enter; embark}.$$

Below is the proof of the regimentation property above with respect to the multi-agent system program with norms that implements the train station with one passenger.

Proof. First, axiom *Ax Frame* using the specification of the *enter* action (with pre-condition $\{\text{-at\_platform}\}$ and post-condition $\{\text{at\_platform}\}$) gives us
(1) $-\text{at\_platform} \wedge -\text{in\_train} \wedge -\text{ticket} \rightarrow$
    $[\text{enter}] \, \text{at\_platform} \wedge -\text{in\_train} \wedge -\text{ticket}$
Moreover, axiom *Ax Frame* using the specification of the *embark* action (with pre-condition $\{\text{at\_platform, -in\_train}\}$ and post-condition $\{\text{-at\_platform, in\_train}\}$) gives us
(2) $\text{at\_platform} \wedge -\text{in\_train} \wedge -\text{ticket} \rightarrow$
    $[\text{embark}] \, -\text{at\_platform} \wedge \text{in\_train} \wedge -\text{ticket}$
Also, axiom *Ax Counts-as* and the specification of the second counts-as rule of the program give us
(3) $\text{in\_train} \wedge -\text{ticket} \rightarrow \text{viol}_\bot$
And axiom *Ax Regiment* together with formula (3) gives us

(4) `in_train` ∧ −`ticket` → ⊥

Now, using PDL axioms together with formula (1), (2), and (4) we get first

(5) −`at_platform` ∧ −`in_train` ∧ −`ticket` → [`enter`][`embark`] ⊥

and thus

(6) −`at_platform`∧−`in_train`∧−`ticket` → [`enter`; `embark`] ⊥. This completes the derivation.

5   Conclusions and Future Work

The paper has proposed a programming language for implementing multi-agent systems with norms. The programming language has been endowed with formal operational semantics, therefore formally grounding the use of certain social notions —eminently the notion of norm, regimentation and enforcement— as explicit programming constructs. A sound and complete logic has then been proposed which can be used for verifying properties of the multi-agent systems with norms implemented in the proposed programming language.

We have already implemented an interpreter for the programming language that facilitates the implementation of multi-agent systems without norms (see http://www.cs.uu.nl/2apl/). Currently, we are working to build an interpreter for the modified programming language. This interpreter can be used to execute programs that implement multi-agent systems with norms. Also, we are working on using the presented logic to devise a semi-automatic proof checker for verification properties of normative multi-agent programs.

We are aware that for a comprehensive treatment of normative multi-agent systems we need to extend our framework in many different ways. Future work aims at extending the programming language with constructs to support the implementation of a broader set of social concepts and structures (e.g., roles, power structure, task delegation, and information flow), and more complex forms of enforcement (e.g., policing agents) and norm types (e.g., norms with deadlines). Another extension of the work is the incorporation of the norm-awareness of agents in the design of the multi-agent system. We also aim at extending the framework to capture the role of norms and sanctions concerning the interaction between individual agents.

The approach in its present form concerns only closed multi-agent systems. Future work will also aim at relaxing this assumption providing similar formal semantics for open multi-agent systems. Finally, we have focused on the so-called 'ought-to-be' norms which pertain to socially preferable states. We intend to extend our programming framework with 'ought-to-do' norms pertaining to socially preferable actions.

References

1. Alechina, N., Dastani, M., Logan, B., Meyer, J.-J.Ch.: A logic of agent programs. In: Proc. AAAI 2007 (2007)
2. Blackburn, P., de Rijke, M., Venema, Y.: Modal Logic. Cambridge University Press, Cambridge (2001)

3. Castelfranchi, C.: Formalizing the informal?: Dynamic social order, bottom-up social control, and spontaneous normative relations. JAL 1(1-2), 47–92 (2004)
4. Dastani, M.: 2apl: a practical agent programming language. International Journal of Autonomous Agents and Multi-Agent Systems 16(3), 214–248 (2008)
5. Dastani, M., Meyer, J.-J.C.: A practical agent programming language. In: Dastani, M.M., El Fallah Seghrouchni, A., Ricci, A., Winikoff, M. (eds.) ProMAS 2007. LNCS (LNAI), vol. 4908, pp. 107–123. Springer, Heidelberg (2008)
6. Dignum, V.: A Model for Organizational Interaction. PhD thesis, Utrecht University, SIKS (2003)
7. Esteva, M., Rodríguez-Aguilar, J.A., Rosell, B., Arcos, J.L.: Ameli: An agent-based middleware for electronic institutions. In: Proc. of AAMAS 2004, New York, US (July 2004)
8. Esteva, M., Rodríguez-Aguilar, J.A., Sierra, C., Garcia, P., Arcos, J.L.: On the formal specification of electronic institutions. In: Sierra, C., Dignum, F.P.M. (eds.) AgentLink 2000. LNCS (LNAI), vol. 1991, pp. 126–147. Springer, Heidelberg (2001)
9. Garcia-Camino, A., Noriega, P., Rodriguez-Aguilar, J.A.: Implementing norms in electronic institutions. In: Proc. of AAMAS 2005, pp. 667–673. ACM, New York (2005)
10. Grossi, D.: Designing Invisible Handcuffs. PhD thesis, Utrecht University, SIKS (2007)
11. Hübner, J.F., Sichman, J.S., Boissier, O.: Moise+: Towards a structural functional and deontic model for mas organization. In: Proc. of AAMAS 2002. ACM, New York (2002)
12. Hubner, J.F., Sichman, J.S., Boissier, O.: Developing organised multiagent systems using the moise+ model: programming issues at the system and agent levels. Int. J. Agent-Oriented Softw. Eng. 1(3/4), 370–395 (2007)
13. Jones, A.J.I., Sergot, M.: On the characterization of law and computer systems. In: Deontic Logic in Computer Science (1993)
14. Kitio, R., Boissier, O., Hübner, J.F., Ricci, A.: Organisational artifacts and agents for open multi-agent organisations: "giving the power back to the agents". In: Sichman, J.S., Padget, J., Ossowski, S., Noriega, P. (eds.) COIN 2007. LNCS (LNAI), vol. 4870, pp. 171–186. Springer, Heidelberg (2008)
15. Plotkin, G.D.: A structural approach to operational semantics. Technical Report DAIMI FN-19, University of Aarhus (1981)
16. Ricci, A., Viroli, M., Omicini, A.: "Give agents their artifacts": The A&A approach for engineering working environments in MAS. In: Proc. of AAMAS 2007, Honolulu, Hawai'i, USA (2007)
17. Searle, J.: The Construction of Social Reality. Free (1995)
18. Silva, V.T.: From the specification to the implementation of norms: an automatic approach to generate rules from norms to govern the behavior of agents. JAAMAS 17(1), 113–155 (2008)
19. Zambonelli, F., Jennings, N., Wooldridge, M.: Developing multiagent systems: the GAIA methodology. ACM Transactions on Software Engineering and Methodology 12(3), 317–370 (2003)

Modal Logics for Preferences and Cooperation: Expressivity and Complexity

Cédric Dégremont and Lena Kurzen[*]

Universiteit van Amsterdam

Abstract. This paper studies expressivity and complexity of normal modal logics for reasoning about cooperation and preferences. We identify a class of local and global notions relevant for reasoning about cooperation of agents that have preferences. Many of these notions correspond to game- and social choice-theoretical concepts. We specify the expressive power required to express these notions by determining whether they are invariant under certain relevant operations on different classes of Kripke models and frames. A large class of known extended modal languages is specified and we show how the chosen notions can be expressed in fragments of this class. To determine how demanding reasoning about cooperation is in terms of computational complexity, we use known complexity results for extended modal logics and obtain for each local notion an upper bound on the complexity of modal logics expressing it.

1 Introduction

Cooperation of agents is a major issue in fields such as computer science, economics and philosophy. The conditions under which coalitions are formed occur in various situations involving multiple agents. A single airline company for instance cannot afford the cost of an airport runway whereas a group of companies can. Generally, agents can form groups in order to share complementary resources or because as a group they can achieve better results than individually. Modal logic (ML) frameworks for reasoning about cooperation mostly focus on what coalitions can achieve. Coalition Logic (**CL**) [1] uses modalities of the form $[C]\phi$ saying that "coalition C has a joint strategy to ensure that ϕ". **CL** has neighborhood semantics but it has been shown how it can be simulated on Kripke models [2].

Another crucial concept for reasoning about interactive situations is that of *preferences*. It also received attention from modal logicians ([3] surveys). Recent works (e.g. [4,5]) propose different mixtures of cooperation and preference logics for reasoning about cooperation. In such logics many concepts from *game theory* (GT) and *social choice theory* (SCT) are commonly encountered. Depending on the situations to be modelled, different bundles of notions are important. Ability

---

[*] Authors are supported by a GLoRiClass fellowship of the EU Commission (Research Training Fellowship MEST-CT-2005-020841) and by the Dutch Organization for Scientic Research (NWO), TACTICS project, grant number 612.000.525, respectively.

J.-J.Ch. Meyer and J.M. Broersen (Eds.): KRAMAS 2008, LNAI 5605, pp. 32–50, 2009.

to express these notions – together with good computational behavior – make a logic appropriate for reasoning about the situations under consideration.

Rather than proposing a new logical framework, with specific expressivity and complexity, we identify how *social choice theory* and *game theory* notions are demanding for MLs in terms of expressivity and complexity. We identify notions relevant for describing interactive situations. Some of them are local, i.e. they are properties of pointed models.

We determine under which operations on models these properties are invariant. Other properties are global, i.e. they are properties of frames. For each of them, we check whether a class of frames having this property is closed under certain operations. We refer to such results as satisfiability invariance and validity closure results respectively.

We also give explicit definability results for them. Given a local property P we give a formula ϕ such that a pointed model $\mathcal{M}, w \Vdash \phi$ iff $\mathcal{M}, w$ has property P. Given a global property Q we give a formula ψ such that a frame $\mathcal{F} \Vdash \phi$ iff $\mathcal{F}$ has property P. We thus identify the natural (extended) modal languages needed depending on the class of frames actually considered and the particular bundle of notions of interest. We draw some consequences about the complexity of reasoning about cooperation using ML.

Our results apply to logics interpreted on Kripke structures using a (preference) relation for each agent and a relation for each coalition. The latter can be interpreted in various ways. The pair (x, y) being in the relation for coalition C can e.g. mean:

- Coalition C considers y as being at least as good as x.
- If the system is in state x, C would choose y as the next state.
- C can submit a request such that if it is the first one received by the server while the state is in x, then the state of the system will change from x to y.
- When the system is in state x, C considers it possible that it is in state y.

Interpreting the relation as the possibility to bring the system into a different state applies to scenarios where agents act sequentially (e.g. with a server treating requests in a "first-come, first-served" manner) rather than simultaneously (as in ATL [6] or **CL**). In special cases - e.g. for turn-based [7,1] frames - the approaches coincide. Still, the two approaches are first of all complementary. Our focus in this paper is on concepts bridging powers and preferences. The same analysis is possible for powers themselves in ATL-style. Both analyses can then be combined in an interesting way. Finally, an important alternative interpretation of the coalition relation is that of group preferences, in which case ATL models can simply be merged with the models we consider. We discuss the possible interpretations of such models in more details in Section 2.

Structure of this Paper. Sect. 2 presents three classes of models of cooperative situations. Sect. 3 introduces local and global notions motivated by ideas from GT and SCT indicating local properties of a system and global properties that characterize classes of frames. Sect. 4 presents a large class of extended modal languages and background invariance results. In Sect. 5, we study the expressivity needed to express the local notions (to define the global properties) by giving invariance results for relevant operations and relations between models (frames). Sect. 6 completes this work by defining the notions in fragments of (extended) modal languages. We give complexity results for model checking and satisfiability for these languages and thereby give upper bounds for the complexity of logics that can express the introduced notions. Sect. 7 concludes.

2 The Models

Our aim is to study how demanding certain GT and SCT concepts are in terms of expressivity and complexity. This depends on the models chosen. We consider three classes of simple models that have many suitable interpretations. This gives our results additional significance. A *frame* refers to the relational part of a model. For simplicity, we introduce models and assume that the domain of the valuation is a countable set of propositional letters PROP and nominals NOM. We focus on model theory and postpone discussion of formal languages to Sect. 4.

Definition 1 (N-LTS). *A N-LTS (Labeled Transition Systems indexed by a finite set of agents* N*) is of the form* $\langle W, \mathsf{N}, \{ \overset{\mathsf{C}}{\to} \mid \mathsf{C} \subseteq \mathsf{N} \}, \{ \leq_i \mid i \in \mathsf{N} \}, V \rangle$, *where* $W \neq \emptyset$, $\mathsf{N} = \{1, \ldots, n\}$ *for some* $n \in \mathbb{N}$, $\overset{\mathsf{C}}{\to} \subseteq W \times W$ *for each* $\mathsf{C} \subseteq \mathsf{N}$, $\leq_j \subseteq W \times W$ *for each* $j \in \mathsf{N}$, *and* $V : \mathsf{PROP} \cup \mathsf{NOM} \to \wp W$, $|V(i)| = 1$ *for each* $i \in \mathsf{NOM}$.

W is the set of states, N a set of agents and $w \overset{\mathsf{C}}{\to} v$ says that coalition C can change the state of the system from w into v. As mentioned, other interpretations are possible. $w \leq_i v$ means that i finds the state v at least as good as w. $w \in V(p)$ means that p is true at w. Preferences are usually assumed to be total pre-orders (TPO). Let TPO-N-LTS denote the class of $\mathsf{N} - \mathsf{LTSs}$ in which for each $i \in \mathsf{N}$, $\leq_i$ is a TPO. We also consider models with strict preferences as explicit primitives.

Definition 2 (S/TPO $-$ N $-$ LTS). *Define* S/TPO $-$ N $-$ LTS *as models of the form* $\langle W, \mathsf{N}, \{ \overset{\mathsf{C}}{\to} \mid \mathsf{C} \subseteq \mathsf{N} \}, \{ \leq_i \mid i \in \mathsf{N} \}, \{ <_i \mid i \in \mathsf{N} \}, V \rangle$, *which extend* TPO $-$ N $-$ LTS *models by an additional relation* $<_i \subseteq W \times W$ *for each* $i \in \mathsf{N}$ *with the constraint that for each* $i \in \mathsf{N}$, $w <_i v$ *iff* $w \leq_i v$ *and* $v \not\leq_i w$.

Depending on the interpretation of $\overset{\mathsf{C}}{\to}$, it can be complemented or replaced by effectivity functions (**CL**) or more generally transition functions as in ATL. In the latter sense, powers of coalitions will in general not reduce to relations on states. We leave an analysis of powers in such settings aside for now. There would be two ways to go: drawing on the model-theory of neighborhood semantics [8] or on a normal simulation of **CL** [2]. Generally, the expressive power might depend on whether coalitional powers are taken as primitives or computed from individual powers.

  In the next section, we identify a list of notions inspired by concepts from game theory and social choice theory for reasoning about cooperative ability and preferences of agents. Then we will determine the expressivity required by certain local and global notions, by giving invariance results for pointed models and closure conditions of classes of frames, respectively. Since we are also interested in the effects of the underlying models on the expressivity required to express the local notions, we will give invariance results with respect to the three different types of models we just introduced.

3 The Notions

Reasoning about cooperative interaction considers what coalitions of agents can achieve and what individuals prefer. Using these elements, more elaborated notions can be built. We consider natural counterparts of SCT and GT notions and are interested both in local notions i.e. properties of a particular state in a particular system, i.e. properties of pointed models $\mathcal{M}, w$, and also in global notions, which are properties of classes of systems. In other words, we are interested in the class of frames a global property characterizes. With respect to content, apart from notions describing only coalitional powers or preferences, we consider stability and effectivity concepts.

Power of Coalitions. We now present some interesting notions about coalitional power. Recall that $w \xrightarrow{C} v$ can e.g. mean "C can achieve v at w".

Local Notions. Interesting properties of coalitional power involve the relation between the powers of different groups ($PowL3$) and the contribution of individuals to a group's power, e.g. an agent is needed to achieve something ($PowL2$).

- $PowL1$. Coalition C can achieve a state where p is true. $\exists x(w \xrightarrow{C} x \wedge P(x))$
- $PowL2$. Only groups with i can achieve p-states. $\bigwedge_{C \subseteq N \setminus i}(\forall x(w \xrightarrow{C} x \Rightarrow \neg P(x)))$
- $PowL3$. Coalition C can force every state that coalition D can force.
  $\forall x(w \xrightarrow{D} x \Rightarrow w \xrightarrow{C} x)$

Global Notions. $PowG1$ says that each coalition can achieve exactly one result. $PowG3$ expresses coalition monotonicity: it says that if a coalition can achieve some result, then so can every superset of that coalition. In many situations, decision making in groups can only be achieved by a majority ($PowG2$). $PowG4$ and $PowG5$ exemplify (mathematically natural) consistency requirements between powers of non-overlapping coalitions.

- $PowG1$. In any state each coalition can achieve exactly one state.
  $\bigwedge_{C \subseteq N} \forall x \exists y (x \xrightarrow{C} y \wedge \forall z(x \xrightarrow{C} z \Rightarrow z = y))$
- $PowG2$. Only coalitions containing a majority of N can achieve something.
  $\forall x(\bigwedge_{C \subseteq N, |C| < \frac{|N|}{2}}(\neg \exists y(x \xrightarrow{C} y)))$
- $PowG3$. Coalition monotonicity, i.e. if for C and D, $C \subseteq D$, then $R_C \subseteq R_D$.
  $\forall x(\bigwedge_{C \subseteq N} \bigwedge_{D \subseteq N, C \subseteq D}(\forall y(x \xrightarrow{C} y \Rightarrow x \xrightarrow{D} y)))$
- $PowG4$. If C can achieve something, then subsets of its complement cannot achieve anything.
  $\forall x \bigwedge_{C \subseteq N}(((\exists y(x \xrightarrow{C} y)) \Rightarrow \bigwedge_{D \subseteq N \setminus C} \neg \exists z(x \xrightarrow{D} z)))$
- $PowG5$. If C can achieve something, then subsets of its complement cannot achieve something C cannot achieve.
  $\forall x \bigwedge_{C \subseteq N}(((\exists y(x \xrightarrow{C} y)) \Rightarrow \bigwedge_{D \subseteq N \setminus C} \forall z(x \xrightarrow{D} z \Rightarrow x \xrightarrow{C} z)))$

Preferences. What do agents prefer? What are suitable global constraints on preferences? $w \leq_i v$ means "i finds v *at least as good* (a.l.a.g.) as w". We write $w <_i v$ for $w \leq_i v \wedge \neg(v \leq_i w)$, meaning that "$i$ *strictly prefers* v over w".

Local Notions. First of all, we can distinguish between strict and nonstrict preferences. The most basic preference relation that we consider is that of being a.l.a.g. We can also look at the relation "at least as bad" (a.l.a.b) ($PrefL4$). Agents' preferences over states can also be seen as being based on preferences over propositions [9]. $PrefL8$ ($PrefL10$) says the truth of a given proposition is a sufficient (necessary) condition for an agent to prefer some state. In what follows, "at least as good" (a.l.a.g) means "at least as good *as the current state*".

- $PrefL1$. There is a state i finds a.l.a.g. where p holds. $\exists x(w \leq_i x \wedge P(x))$
- $PrefL2$. There is a p-state that i strictly prefers. $\exists x(w <_i x \wedge P(x))$
- $PrefL3$.There is a state that all agents find a.l.a.g and that at least one strictly prefers. $\exists x(\bigwedge_{i \in \mathbb{N}}(w \leq_i x) \wedge \bigvee_{j \in \mathbb{N}} w <_j x)$
- $PrefL4$. There is a state that i finds a.l.a.b. where p holds. $\exists x(x \leq_i w \wedge P(x))$
- $PrefL5$. There is a state that i finds strictly worse where p is true. $\exists x(x <_i w \wedge P(x))$
- $PrefL6$. i finds a state a.l.a.g. a the current one iff j does. $\forall x(w \leq_i x \leftrightarrow w \leq_j x)$
- $PrefL7$. There is a state only i finds a.l.a.g. $\exists x(w \leq_i x \wedge \bigwedge_{j \in \mathbb{N} \setminus \{i\}} \neg(w \leq_j x))$
- $PrefL8$. i finds every p-state a.l.a.g. $\forall x(P(x) \Rightarrow w \leq_i x)$
- $PrefL9$. i strictly prefers every p-state. $\forall x(P(x) \Rightarrow w <_i x)$
- $PrefL10$. i considers only p-states to be a.l.a.g. $\forall x(w \leq_i x \Rightarrow P(x))$
- $PrefL11$. i strictly prefers only p-states. $\forall x(w <_i x \Rightarrow P(x))$

Global Notions. Capturing the intuitive idea of preferences requires several conditions for the preference relation: reflexivity, transitivity and completeness (trichotomy for strict preferences). Sometimes, it can also be appropriate to say that for each alternative there is exactly one that is at least as good ($PrefG8$).

- $PrefG1$. "at least as good as" is reflexive. $\forall x(\bigwedge_{i \in \mathbb{N}}(x \leq_i x))$
- $PrefG2$. "at least as good as" is transitive.
  $\forall x \forall y \forall z(\bigwedge_{i \in \mathbb{N}}((x \leq_i y \wedge y \leq_i z) \Rightarrow x \leq_i z))$
- $PrefG3$. "at least as good as" is complete.
  $\forall x \forall y(\bigwedge_{i \in \mathbb{N}}(x \leq_i y \vee y \leq_i x))$
- $PrefG4$. "at least as good as" is a total pre-order.
  (Conjunction of the two previous formulas.)
- $PrefG5$. "strictly better than" is transitive.
  $\forall x \forall y \forall z((\bigwedge_{i \in \mathbb{N}}(x <_i y \wedge y <_i z) \Rightarrow x <_i z)))$
- $PrefG6$. "strictly better than" is trichotomous.
  $\forall x \forall y(\bigwedge_{i \in \mathbb{N}}(x <_i y \vee y <_i x \vee x = y))$
- $PrefG7$. "strictly better than" a strict total order.
  (Conjunction of the previous two formulas.)
- $PrefG8$. Determinacy for "at least as good as", i.e. exactly one successor.
  $\forall x(\bigwedge_{i \in \mathbb{N}}(\exists y(w \leq_i y \wedge \forall z(x \leq_i z \Rightarrow z = y))))$

So far, we focussed on preferences of individuals. A natural question in SCT is how to aggregate individual preferences into group preferences. We can address this question by interpreting $\overset{C}{\to}$ as a preference relation for each $C \subseteq N$.

- $PrefG9$. C finds a state a.l.a.g. as the current one iff all its members do.
  $$\forall x \forall y (\bigwedge_{C \subseteq N} (x \overset{C}{\to} y \leftrightarrow \bigwedge_{i \in C} x \leq_i y))$$
- $PrefG10$. C finds a state at least as good as the current one iff at least one member does.
  $$\forall x \forall y (\bigwedge_{C \subseteq N} (x \overset{C}{\to} y \leftrightarrow \bigvee_{i \in C} x \leq_i y))$$
- $PrefG11$. C finds a state a.l.a.g. as the current one iff most members do.
  $$\forall x \forall y (\bigwedge_{C \subseteq N} (x \overset{C}{\to} y \leftrightarrow \bigvee_{D \subseteq C, |D| > \frac{|C|}{2}} (\bigwedge_{i \in D} x \leq_i y)))$$

Combining preceding concepts. We start with the conceptually and histori- cally important SCT notion of a *dictator*. d is a dictator if the group's preferences mimic d's preferences. Interpreting $\overset{C}{\to}$ as achievement relation, we get an even stronger notion: groups can only *do* what d likes. A *local* dictator is a dictator who controls one state in the system, and a *dictator* controls all states.

Definition 3 (Local Dictatorship). *i is a weak (strong) local dictator at w iff any group prefers v at w only if for i, v is a.l.a.g. as (strictly better than) w.*

We now introduce combinations of powers and preferences. The first notion says that coalition C can do something useful for i (in some cases giving i an incentive to join) and the third notion characterizes situations in which a unanimously desired state remains unachievable. We start with **Local Notions**.

- $PPL1$. C can achieve a state that i finds at least as good as the current one.
  $\exists x (w \overset{C}{\to} x \wedge w \leq_i x)$
- $PPL2$. C can achieve a state that all $i \in D$ find a.l.a.g. as the current one.
  $\exists x (w \overset{C}{\to} x \wedge \bigwedge_{i \in D} w \leq_i x)$
- $PPL3$. There is a state that all agents prefers but no coalition can achieve it. $\exists x ((\bigwedge_{i \in N} w \leq_i x) \wedge \bigwedge_{C \subseteq N} \neg(w \overset{C}{\to} x))$
- $PPL4$. C can achieve all states that agent i finds a.l.a.g. as the current one.
  $\forall x (w \leq_i x \Rightarrow w \overset{C}{\to} x)$
- $PPL5$. C can achieve all states that i strictly prefers. $\forall x (w <_i x \Rightarrow w \overset{C}{\to} x)$
- $PPL6$. i is a weak local dictator. $\forall x (\bigwedge_{C \subseteq N} (w \overset{C}{\to} x \Rightarrow w \leq_i x))$
- $PPL7$. i is a strong local dictator. $\forall x (\bigwedge_{C \subseteq N} (w \overset{C}{\to} x \Rightarrow w <_i x))$

Global Notions. $PPG1$ is a natural constraint on coalitional power: a group can achieve a state iff it is good for all members - otherwise they would not take part in the collective action. $PPG3$ is a condition of Arrow's impossibility the- orem. $PPG4$ reflects individual rationality: don't join a group if you don't gain anything. It can be generalized to every sub-coalition or weakened to "not joining if you lose something" (cf. core of a coalitional game [10] (Def. 268.3)). $PPG5$ applies to systems where an agent is indispensable to achieve anything: a unique capitalist in a production economy or a unique server are typical examples.

- $PPG1$. Coalitions can only achieve states that all its members consider at least as good as the current one. $\forall x \forall y \bigwedge_{C \subseteq N}(x \xrightarrow{C} y \Rightarrow \bigwedge_{i \in C}(x \leq_i y))$
- $PPG2$. One agent is a weak local dictator in every state (*dictator*). $\bigvee_{i \in N} \forall x \forall y(x \xrightarrow{C} y \Rightarrow x \leq_i y)$
- $PPG3$. There is no *dictator*. $\neg(\bigvee_{i \in N} \forall x \forall y(x \xrightarrow{C} y \Rightarrow x \leq_i y))$
- $PPG4$. If i can achieve some state i strictly prefers then for any C containing i: if $C \setminus i$ cannot achieve some state but C can, then i strictly prefers that state. $\bigwedge_{i \in N} \forall x(\exists y(x \xrightarrow{\{i\}} y \wedge x <_i y) \Rightarrow \bigwedge_{C \subseteq N, i \in C}(\forall z(x \xrightarrow{C} z \wedge \neg(x \xrightarrow{C \setminus \{i\}} z)) \Rightarrow x <_i z))$
- $PPG5$. Only groups with i can achieve something. $\forall x \bigwedge_{C \subseteq N \setminus \{i\}} \neg \exists y(x \xrightarrow{C} y)$
- $PPG6$. In all states, there is an i such that groups with i can achieve exactly the states as they can without i. $\forall x(\bigvee_{i \in N} \bigwedge_{C \subseteq N, i \in C} \forall y(x \xrightarrow{C} y \leftrightarrow x \xrightarrow{C \setminus \{i\}} y))$
- $PPG7$. For any agent, there is some state in which coalitions not containing this agent cannot achieve any state. $\bigwedge_{i \in N} \exists x(\bigwedge_{C \subseteq N, i \in C} \neg \exists y(x \xrightarrow{C} y))$

Efficiency and Stability Notions. In our setting, it is natural to interpret the state space as possible social states or allocations of goods. A criterion from welfare economics to distinguish "good" from "bad" states is that of *efficiency*: if we can change the allocation or social state and make an agent happier without making anyone less happy then we are using resources more efficiently and it is socially desirable to do so. E.g. $Pref L3$ in this respect means that the current state is not efficient: there is a state that is a *Pareto-improvement* of it. Importing the notion of Pareto-efficiency into our framework is straightforward.

Definition 4 (Pareto-efficiency). *A state is weakly (strongly) Pareto-efficient iff there is no state that everyone strictly prefers (finds a.l.a.g). A state is Pareto-efficient iff there is no state such that everyone considers it to be at least as good and at least one agent thinks that it is strictly better.*

GT equilibrium concepts characterize stable states: given what others are doing, I don't have an incentive to do something that makes us leave this stable state. Generalizing, a system is in a stable state if nobody has an incentive to change its current state. We can think of strategy profiles in a strategic game as assigning roles to the agents. Two profiles $x = (s^*_{-i}, s^*_i), y = (s^*_{-i}, s'_i)$ are related by $\xrightarrow{\{i\}}$ iff i can unilaterally change role (strategy) to s'_i in the next round of the game. E.g. the stability of a state where an agent provides the public good on his own depends on whether he cares enough about it to provide it on his own. A state is stable iff there is no strictly preferred state that an agent can achieve alone. Since the idea relates to *Nash* equilibria (see [10]), we use the names *Nash-stability*, and *Nash-cooperation stability* for its group version.

Definition 5 (Nash-stability). *A state is (strongly) Nash-stable iff there is no state that an agent i strictly prefers (finds a.l.a.g.) and that i can achieve alone. It is (strongly) Nash-cooperation stable iff there is no state v and coalition such that C that every $i \in C$ strictly prefers v (finds v a.l.a.g.) and C can achieve v.*

Local Notions

- $EF1$.The current state is weakly *Pareto*-efficient. $\neg\exists x(\bigwedge_{i\in\mathbb{N}}(w <_i x))$
- $EF2$. The current state is *Pareto*-efficient. $\neg\exists x((\bigwedge_{i\in\mathbb{N}} w \leq_i x) \wedge \bigvee_{j\in\mathbb{N}} w <_i x)$
- $EF3$. The current state is strongly *Pareto*-efficient. $\neg\exists x(\bigwedge_{i\in\mathbb{N}} w \leq_i x)$
- $ST1$. The current state is *Nash* stable. $\neg\exists x(\bigvee_{i\in\mathbb{N}}(w \overset{\{i\}}{\to} x \wedge w <_i x))$
- $ST2$. The current state is strongly *Nash* stable. $\neg\exists x(\bigvee_{i\in\mathbb{N}}(w \overset{\{i\}}{\to} x \wedge w \leq_i x))$
- $ST3$. The current state is *Nash*-cooperation stable.
  $\neg\exists x(\bigvee_{C\subseteq\mathbb{N}}(w \overset{C}{\to} x \wedge \bigwedge_{i\in C} w <_i x))$
- $ST4$. The current state is strongly *Nash*-cooperation stable.
  $\neg\exists x(\bigvee_{C\subseteq\mathbb{N}}(w \overset{C}{\to} x \wedge \bigwedge_{i\in C} w \leq_i x))$

4   Modal Languages and Their Expressivity

As will be clear from invariance results of next sections, Basic Modal Language will generally be too weak for reasoning about cooperation. However, any notion expressible in the FO correspondence language is expressible in the hybrid language $\mathcal{H}(\mathrm{E}, @, \downarrow)$ [11]. Amongst temporal logics, boolean modal logics and the various hybrid logics, there are well-understood fragments. We introduce all these **Extended Modal Languages** at once as a "super" logic.

Syntax. The syntax of this "super" logic is recursively defined as follows:
$$\alpha ::= \quad \leq_j \mid \mathsf{C} \mid v \mid \alpha^{-1} \mid ?\phi \mid \alpha;\alpha \mid \alpha \cup \alpha \mid \alpha \cap \alpha \mid \overline{\alpha}$$
$$\phi ::= \quad p \mid i \mid x \mid \neg\phi \mid \phi \wedge \phi \mid \langle\alpha\rangle\phi \mid \mathrm{E}\phi \mid @_i\phi \mid @_x\phi \mid \downarrow x.\phi \mid [\![\,\alpha\,]\!]\phi \mid$$
where $j \in \mathbb{N}$, $\mathsf{C} \in \wp(\mathbb{N}) - \{\emptyset\}$, p ranges over PROP, i ranges over NOM and $x \in$ SVAR, for SVAR being a countable set of variables.

Semantics. Valuation maps propositional letters to subsets of the domain and nominals to singleton subsets. Given a $\mathbb{N}-$LTS, a program α is interpreted as a relation as indicated on the left. Formulas are interpreted together with an assignment $g :$ SVAR $\to W$ as indicated (mostly) on the right. We skip booleans.

$\mathcal{M}, w, g \Vdash i$ iff $w \in V(i)$		$\mathcal{M}, w, g \Vdash x$	iff $w = g(x)$	
$R_{\leq_i}$	$= \leq_i$	$\mathcal{M}, w, g \Vdash \langle\alpha\rangle\phi$	iff $\exists v : wR_\alpha v$ and $\mathcal{M}, v, g \Vdash \phi$	
R_C	$= \overset{C}{\to}$	$\mathcal{M}, w, g, \Vdash \mathrm{E}\phi$	iff $\exists v \in W$ $\mathcal{M}, v, g \Vdash \phi$	
$R_{\beta^{-1}}$	$= \{(v,w)\mid wR_\beta v\}$	$\mathcal{M}, w, g, \Vdash @_i\phi$	iff $\mathcal{M}, v, g \Vdash \phi$ where $V(i) = \{v\}$	
$R_{\beta\cup\gamma}$	$= R_\beta \cup R_\gamma$	$\mathcal{M}, w, g, \Vdash @_x\phi$	iff $\mathcal{M}, g(x), g \Vdash \phi$	
$R_{\beta\cap\gamma}$	$= R_\beta \cap R_\gamma$	$\mathcal{M}, w, g, \Vdash \downarrow x.\phi$	iff $\mathcal{M}, w, g[x := w] \Vdash \phi$	
$R_{\overline{\beta}}$	$= (W \times W) - R_\beta$	$\mathcal{M}, w, g \Vdash [\![\,\alpha\,]\!]\phi$ iff $wR_\alpha v$ whenever $\mathcal{M}, v, g \Vdash \phi$		

Expressivity. The least expressive modal language we consider is $\mathcal{L}(\mathbb{N})$, which is of similarity type $\langle(\mathsf{C})_{\mathsf{C}\subseteq\mathbb{N}}, (\leq_i)_{i\in\mathbb{N}}\rangle$. Its natural extensions go along two lines: adding program constructs and new operators. $\mathcal{L}(\mathbb{N}, \cap, i)$ e.g. refers to the logic with language: $\alpha ::= \leq_j \mid \mathsf{C} \mid \alpha \cap \alpha \quad \phi ::= p \mid i \mid \neg\phi \mid \phi \wedge \phi \mid \langle\alpha\rangle\phi$. As language inclusion implies expressivity inclusion (indicated by "$\leq$"), we only indicate (some) non-obvious facts of inclusions in this space of modal languages.

Fact 1. $\mathcal{L}(\mathtt{N}, \cup, \; ; \, , ?) \leq \mathcal{L}(\mathtt{N}).$

Proof. By the facts that $\mathcal{M}, w, g \Vdash \langle \alpha \cup \beta \rangle \phi$ iff $\mathcal{M}, w, g \Vdash \langle \alpha \rangle \phi \vee \langle \beta \rangle \phi$, $\mathcal{M}, w, g \Vdash \langle \alpha ; \beta \rangle \phi$ iff $\mathcal{M}, w, g \Vdash \langle \alpha \rangle \langle \beta \rangle \phi$ and by $\mathcal{M}, w, g \Vdash \langle ? \psi \rangle \phi$ iff $\mathcal{M}, w, g \Vdash \psi \wedge \phi$.

Fact 2. $\mathcal{L}(\mathtt{N}, @, i) \leq \mathcal{L}(\mathtt{N}, \mathtt{E}, i).$

Proof. By the fact that $\mathcal{M}, w, g \Vdash @_i \phi$ iff $\mathcal{M}, w, g \Vdash \mathtt{E}(i \wedge \phi)$.

Fact 3. $\mathcal{L}(\mathtt{N}, \cap) \leq \mathcal{L}(\mathtt{N}, \downarrow, @, x).$

Proof. By the fact that $\mathcal{M}, w, g \Vdash \langle \alpha \cap \beta \rangle \phi$ iff $\mathcal{M}, w, g \Vdash \downarrow x. \langle \alpha \rangle (\downarrow y. \phi \wedge @_x \langle \beta \rangle y).$

Fact 4. $\mathcal{L}(\mathtt{N}, [\![\;]\!]) \leq \mathcal{L}(\mathtt{N}, \text{-}).$

Proof. By the fact that $\mathcal{M}, w, g \Vdash [\![\; \alpha \;]\!] \phi$ iff $\mathcal{M}, w, g \Vdash [\overline{\alpha}] \neg \phi$.

Fact 5. $\mathcal{L}(\mathtt{N}, \text{-}) \leq \mathcal{L}(\mathtt{N}, \downarrow, \mathtt{E}, x).$

Proof. By the fact that $\mathcal{M}, w, g \Vdash \langle \overline{\alpha} \rangle \phi$ iff $\mathcal{M}, w, g \Vdash \downarrow x. \mathtt{E} \downarrow y. (\phi \wedge \neg \mathtt{E}(x \wedge \langle \alpha \rangle y)).$

Fact 6. $\mathcal{L}(\mathtt{N}, \; ^{-1}) \leq \mathcal{L}(\mathtt{N}, \downarrow, \mathtt{E}, x).$

Proof. By the fact that $\mathcal{M}, w, g \Vdash \langle \alpha^{-1} \rangle \phi$ iff $\mathcal{M}, w, g \Vdash \downarrow x. \mathtt{E}(\phi \wedge \langle \alpha \rangle x).$

Fact 7. $\mathcal{L}(\mathtt{N}, \mathtt{E}) \leq \mathcal{L}(\mathtt{N}, \text{-}).$

Proof. By the fact that $\mathcal{M}, w, g \Vdash \mathtt{E} \phi$ iff $\mathcal{M}, w, g \Vdash \langle \alpha \rangle \phi \vee \langle \overline{\alpha} \rangle \phi.$

Expressivity of MLs is usually characterized by invariance results. Definitions and background results follow. We first introduce some relations between models. Let τ be a finite modal similarity type with only binary relations. Let $\mathcal{M} = \langle W, (R_k)_{k \in \tau}, V \rangle$ and $\mathcal{M}' = \langle W', (R'_k)_{k \in \tau}, V' \rangle$ be models of similarity type τ.

Definition 6 (Bisimulations). *A bisimulation between $\mathcal{M}$ and $\mathcal{M}'$ is a non-empty binary relation $Z \subseteq W \times W'$ fulfilling the following conditions:*

AtomicHarmony *For every $p \in$* PROP, *wZw' implies $w \in V(p)$ iff $w' \in V'(p)$.*

Forth $\forall k \in \tau$, *if wZw' & $R_k wv$ then $\exists v' \in W'$ s.t. $R'_k w'v'$ & vZv'.*

Back $\forall k \in \tau$, *if wZw' & $R'_k w'v'$ then $\exists v \in W$ s.t. $R_k wv$ & vZv'.*

In a nutshell, $\cap$-Bisimulations (resp. CBisimulations) require that Back and Forth also hold for the intersection (resp. the converse) of the relations. $\mathcal{H}$-Bisimulations extend AtomicHarmony to nominals. TBisimulations ($\mathcal{H}(@)$-bisimulations) are total [1] bisimulations (resp. total $\mathcal{H}$-Bisimulations). $\mathcal{H}(\mathtt{E})$-Bisimulations are $\mathcal{H}$-Bisimulations matching states "with the same name". See [11] for details. We now define bounded morphisms, generated subframes and disjoint unions.

---

[1] $Z \subseteq W \times W'$ is *total* iff $\forall w \in W \; \exists w' \in W' \; wZw'$ & $\forall w' \in W' \; \exists w \in W \; wZw'$.

Definition 7 (BM). $f : W \to W'$ *is a bounded morphism from* $\mathcal{M}$ *to* $\mathcal{M}'$ *iff:*
AtomicHarmony   *For every* $p \in$ PROP, $w \in V(p)$ *iff* $f(w) \in V'(p)$.
R − homomorphism $\forall k \in \tau$, *if* $R_k wv$ *then* $R' f(w) f(v)$.
Back $\forall k \in \tau$, *if* $R'_k f(w) v'$ *then* $\exists v \in W$ *s.t.* $f(v) = v'$ *and* $R_k wv$.

Definition 8 (Generated Submodel). *We say that that* $\mathcal{M}'$ *is a generated submodel (GSM) of* $\mathcal{M}$ *iff* $W' \subseteq W$, $\forall k \in \tau$, $R'_k = R_k \cap (W' \times W')$, $\forall p \in$ PROP , $V'(p) = V(p) \cap (W' \times W')$ *and if* $w \in W'$ *and* Rwv *then* $v \in W'$.

Definition 9 (Disjoint Unions). *Let* $(\mathcal{M}_j)_{j \in J}$ *be a collection of models with disjoint domains. Define their disjoint union* $\biguplus_j \mathcal{M}_j = \langle W, R, V \rangle$ *as the union of their domains and relations, and define for each* $p \in$ PROP, $V(p) := \bigcup_j V_j(p)$.

Definition 10 (Invariance). *A property of pointed models* $\Phi(X, y)$ *is invariant under* λ*-Bisimulations iff whenever there exists a* λ*-bisimulation* Z *between* $\mathcal{M}$ *and* $\mathcal{M}'$ *such that* $(w, w') \in Z$, *then* $\Phi(\mathcal{M}, w')$ *holds iff* $\Phi(\mathcal{M}', w')$ *holds. Invariance for other operations is defined similarly.*

We now consider closure conditions. First, we consider bounded morphic images (BMI) of frames. BM on frames are obtained by dropping AtomicHarmony in Def. 7. A class of frames is closed under BMI iff it is closed under *surjective* BM. Next, we consider closure under generated subframes (GSF) – the frame-analogue to GSM (cf. Def. 8). We also check if properties *reflect* GSF. A property ϕ reflects GSF if whenever for every frame $\mathcal{F}$, it holds that every GSF of $\mathcal{F}$ has property ϕ, then so does $\mathcal{F}$. We also consider closure under taking disjoint unions (DU) of frames, which are defined in the obvious way. Moreover, we look at closure under images of bisimulation systems [11], which are families of partial isomorphisms.

Definition 11 (Bisimulation System). *A bisimulation system from a frame* $\mathcal{F}$ *to a frame* $\mathcal{F}'$ *is a function* $\mathcal{Z} : \wp W' \to \wp (W \times W')$ *that assigns to each* $Y \subseteq W'$ *a total bisimulation* $\mathcal{Z}(Y) \subseteq W \times W'$ *such that for each* $y \in Y$:
  1. There is exactly one $w \in W$ *such that* $(w, y) \in \mathcal{Z}(Y)$.
  2. If $(w, y), (w, w') \in \mathcal{Z}(Y)$, *then* $w' = y$.

Background results. We indicate three classical characterization results. For details see [12,11]. Let $\phi(x)$ be a formula of the FO correspondence language with at most one free variable. [13] proved that $\phi(x)$ is invariant under bisimulations iff $\phi(x)$ is equivalent to the standard translation of a modal formula. While [14,15] proved that $\phi(x)$ is invariant under taking generated submodels iff $\phi(x)$ is equivalent to the standard translation of a formula of $\mathcal{L}(\mathbb{N}, \downarrow, @, x)$. On the level of frames [16] proved that a FO definable class of frames is modally definable iff it is closed under taking BMI, GSF, disjoint unions and reflects ultrafilter extensions.

The reader might now like to see immediately how the notions can be defined in extended modal languages and go directly to Sect. 6. Of course, the choice of the languages is only justified once we have determined the required expressive power both to express the local notions and to define the class of frames corresponding to the global ones. Thus we start by doing so in the next section.

5   Invariance and Closure Results

We start with satisfiability invariance results for the classes of pointed models defined in Sect. 2. Then we turn to closure results for classes of frames defined by global notions. A "Y" in a cell means that the row notion is invariant under the column operation. The number in the columns refer to representative proofs for these results found below the tables. They will give the reader a concrete idea of the meaning of these results. More of them in our technical report [17].

Overview of the Results for the General Case

	Bis	CBis	∩-Bis	TBis	$\mathcal{H}$-Bis	$\mathcal{H}(@)$-Bis	$\mathcal{H}(E)$-Bis	BM	GSM	DU
$[PowL1]$	Y	Y	Y	Y	Y	Y	Y	Y	Y	Y
$[PowL2]$	Y	Y	Y	Y	Y	Y	Y	Y	Y	Y
$[PowL3]$	N	N	N	N	N	N	N	N	Y	Y
$[PrefL1]$	Y	Y	Y	Y	Y	Y	Y	Y	Y	Y
$[PrefL2]$	N	N	N	N	N	N	N	N	Y	Y
$[PrefL3]$	N	N	N	N	N	N	N	N	Y	Y
$[PrefL4]$	N	Y	N	N	N	N	N	N	N(2)	Y
$[PrefL5]$	N	N	N	N	N	N	N	N	N	Y
$[PrefL6]$	N	N	N	N	N	N	N	N	Y	Y
$[PrefL7]$	N	N	N	N	N	N	N	N	Y	Y
$[PrefL8]$	N	N	N	N	N	N	N	N	N	N
$[PrefL9]$	N	N	N	N	N	N	N	N	N	N
$[PrefL10]$	Y	Y	Y	Y	Y	Y	Y	Y	Y	Y
$[PrefL11]$	N	N	N	N	N	N	N	N	Y	Y
$[PPL1]$	N	N	Y	N	N	N	N	N	Y	Y
$[PPL2]$	N	N	Y	N	N	N	N	N	Y	Y
$[PPL3]$	N	N	N	N	N	N	N	N	Y	Y
$[PPL4]$	N	N	N	N	N	N	N	N	Y	Y
$[PPL5]$	N	N	N	N	N	N	N	N	Y	Y
$[PPL6]$	N	N	N	N	N	N	N	N	Y	Y
$[PPL7]$	N	N	N	N	N	N	N	N	Y	Y
$[EF1]$	N	N	N	N	N	N	N	N	Y	Y
$[EF2]$	N	N	N	N	N	N	N	N	Y	Y
$[EF3]$	N	N	Y	N	N	N	N	N	Y	Y
$[ST1]$	N	N	N	N	N	N	N	N	Y	Y
$[ST2]$	N	N	Y	N	N	N	N	N	Y	Y
$[ST3]$	N	N	N (1)	N	N	N	N	N	Y	Y
$[ST4]$	N	N	Y	N	N	N	N	N	Y	Y

Comments. Most of our notions are not bisimulation-invariant. The basic modal language [2] is thus not expressive enough to describe our local notions (without restrictions on the class of frames). Invariance under BM often fails; some failures are due to intersections of relations, but as ∩-Bis also fails often, this cannot be the only reason. By contrast, invariance under GSM generally holds; it fails for properties with backward looking features. This is good news for expres-

---

[2] of similarity type $\langle \{ \stackrel{c}{\rightarrow} \mid C \subseteq \mathtt{N} \}, \{ \leq_i \mid i \in \mathtt{N} \} \rangle$

sivity: we can expect definability in the hybrid language with $\downarrow$-binder. [3] But not for computability, since the satisfiability problem of the bounded fragment is undecidable. Finally, the results are the same for hybrid and basic bisimulations. No suprise: roughly speaking, at the level of local satisfaction, to exploit the expressive power of nominals, the notions would have to refer explicitly to some state. Here are two representative results.

Representative Proofs for the General Case

Proposition 1. *On* $\mathsf{N-LTS}$, *ST3 is not invariant under* $\cap$-*bisimulation.*

Proof. Let $\mathcal{M} = \langle \{w,v\}, \{1,2\}, \{\overset{C}{\to} \mid C \subseteq \{1,2\}\}, \{\leq_1, \leq_2\}, V\rangle$, where $w \overset{\{1,2\}}{\to} v, w \leq_1 v, v \leq_1 w, w \leq_2 v, V(p) = \{w,v\}$. Let $\mathcal{M}' = \langle \{s,t,u\}, \{1,2\}, \{\overset{C}{\to}' \mid C \subseteq \{1,2\}\}, \{\leq_1', \leq_2'\}, V'\rangle$, where $s \overset{\{1,2\}}{\to}' t, u \overset{\{1,2\}}{\to}' t, s \leq_1' t, u \leq_1' t, t \leq_1' u, s \leq_2' t, u \leq_2' t, V'(p) = \{s,t,u\}$. Then, $\mathcal{M}, w \Vdash ST3$ and $\mathcal{M}', s \nVdash ST3$ because $s \overset{\{1,2\}}{\to}' t$ and $s <_1' t, s <_2' t$. Moreover, $Z = \{(w,s),(w,u),(v,t)\}$ is a $\cap$-bisimulation. □

Proposition 2. *On* $\mathsf{N-LTS}$, *PrefL4 is not invariant under* GSM.

Proof. Let $\mathcal{M} = \langle \{w,v\}, \{1\}, \{\overset{C}{\to} \mid C \subseteq \{1\}\}, \{\leq_1\}, V\rangle$, where $\overset{\{1\}}{\to} = \emptyset$, $v \leq_1 w, V(p) = \{v\}$. Then, $\mathcal{M}, w \Vdash PrefL4$ because $v \leq_1 w$ and $v \in V(p)$. But for the submodel $\mathcal{M}'$ generated by $\{w\}$, $\mathcal{M}', w \nVdash PrefL4$ since v is not contained in $\mathcal{M}'$. □

Results Overview for the Total Pre-orders (TPO) Case. This table shows rows that differ from the general case. Entries that differ are in boldface.

	Bis	CBis	∩-Bis	TBis	$\mathcal{H}$-Bis	$\mathcal{H}$(@)-Bis	$\mathcal{H}$(E)-Bis	BM	GSM	DU
[PrefL8]	N	N	N	N	N	N	N	N	N	**Y***
[PrefL9]	N	**Y** (3)	N	N	N	N	N	N	N	**Y***
[EF1]	N	N	N	N	N	N	N	N	Y	**Y***

Proposition 3. *On* $\mathsf{TPO-N-LTS}$, *PrefL9 is invariant under* $CBis$.

Proof. Let $\mathcal{M}, w$ and $\mathcal{M}', w'$ be two pointed $\mathsf{TPO-N-LTS}$ such that there exists a C-Bisimulation Z between $\mathcal{M}$ and $\mathcal{M}'$ such that $(w,w') \in Z$. But now assume that $\mathcal{M}, w \nVdash PrefL9$. It follows that there exists some $t \in W$ such that $t \in V^{\mathcal{M}}(p)$ but $w \nless_i^{\mathcal{M}} t$. But then by totality of $\leq_i^{\mathcal{M}}$ we have $t \leq_i^{\mathcal{M}} w$, i.e. $w \geq_i^{\mathcal{M}} t$. But then by C-Bisimulation there exists some state $t' \in W'$ such that $(t,t') \in Z$ and therefore $t' \in V^{\mathcal{M}'}(p)$ and $w' \geq_i^{\mathcal{M}'} t'$ and thus $w' \nless_i^{\mathcal{M}'} t'$. By definition we have thus $\mathcal{M}', w' \nVdash PrefL9$. The other direction is symmetrical.

Comments. Except for disjoint union (DU), the restriction to the TPO case brings only slight benefits. * marks trivial invariance: the only DU of models that is complete is the trivial one: mapping a model to itself.

---

[3] [15,14] have proved that all notions definable in the first-order correspondence language that are invariant under GSM are equivalent to a formula of the bounded fragment, i.e. of the hybrid language with $\downarrow$-binder (which are notational variants).

Overview of the Results for the TPO Case with Strict Preferences. The following table contains the rows that differ from the ones in the table for total preorders without strict preference relation.

	Bis	CBis	∩-Bis	TBis	$\mathcal{H}$-Bis	$\mathcal{H}(@)$-Bis	$\mathcal{H}(E)$-Bis	BM	GSM	DU
[PrefL2]	Y	Y	Y	Y	Y	Y	Y	Y	Y	Y
[PrefL3]	N	N	N	N	N	N	N	Y	Y	Y
[PrefL5]	N	Y	N	N	N	N	N	N	N	Y
[PrefL6]	N	N	N	N	N	N	N	Y	Y	Y
[PrefL7]	N	N	N	N	N	N	N	Y	Y	Y
[PrefL8]	N	Y	N	N	N	N	N	N	N	Y
[PrefL11]	Y	Y	Y	Y	Y	Y	Y	Y (4)	Y	Y
[PPL7]	N	N	N	N	N	N	N	Y	Y	Y
[EF1]	N	N	Y	N	N	N	N	Y	Y	Y
[EF2]	N	N	Y	N	N	N	N	Y	Y	Y
[ST1]	N	N	Y	N	N	N	N	Y (5)	Y	Y
[ST3]	N	N	Y	N	N	N	N	Y	Y	Y

Proposition 4. *On* S/TPO-N-LTS, *$PrefL11$ is invariant under* BM.

Proof. Let $\mathcal{M}$ and $\mathcal{M}'$ be two S/TPO − N − LTS and assume that f is a bounded morphism from $\mathcal{M}$ to $\mathcal{M}'$. Assume that the property $PrefL11$ does not hold for $\mathcal{M}, w$, i.e. there is a state $v \in Dom(\mathcal{M})$ such that $w <_i^{\mathcal{M}} v$ and $v \notin V^{\mathcal{M}}(p)$. But then by R − homomorphism, we have $f(w) <_i^{\mathcal{M}'} f(v)$ and by AtomicHarmony, $f(v) \notin V^{\mathcal{M}'}(p)$, and thus $PrefL11$ does not hold for $\mathcal{M}', f(w)$. For the other direction, assume that $PrefL11$ is not satisfied at $\mathcal{M}', f(w)$. It follows that there is a state $v' \in Dom(\mathcal{M}')$ such that $f(w) <_i^{\mathcal{M}'} v'$ and $v' \notin V^{\mathcal{M}'}(p)$ but then by Back there is a state $v \in Dom(\mathcal{M})$ such that $f(v) = v'$ and $w <_i^{\mathcal{M}} v$. But by AtomicHarmony, $v \notin V^{\mathcal{M}}(p)$ and thus $PrefL11$ is not satisfied at $\mathcal{M}, w$, concluding our proof.

Proposition 5. *On* S/TPO-N-LTS, *$ST1$ is invariant under* BM.

Proof. One direction follows directly from R − homomorphism. For the other direction, assume that $PrefL11$ is not satisfied at $\mathcal{M}', f(w)$. It follows that there is a state $v' \in Dom(\mathcal{M}')$ such that $f(w) \overset{\{i\}}{\to} v'$ and $f(w) <_i^{\mathcal{M}'} v'$ (a). But then by Back there is a state $v \in Dom(\mathcal{M})$ such that $f(v) = v'$ and $w \overset{\{i\}}{\to} v$ (b). We are in one of two cases. Case 1: $v \not\leq_i^{\mathcal{M}} w$ (c). But then by totality we have $w \leq_i^{\mathcal{M}} v$ (d). But it follows from (b), (c), (d) that $PrefL11$ is not satisfied at $\mathcal{M}, w$. Now assume for contradiction that we are in the other case. Case 2: $v \leq_i^{\mathcal{M}} w$ (e). But then by R − homomorphism, we have $f(v) \leq_i^{\mathcal{M}'} f(w)$ contradicting the assumption that (a) and concluding our proof.

Comments. The failures of invariance under GSM are still present, reflecting the fact that we do not have converse relations. By contrast, $PrefL11$ and $PrefL2$ are now invariant under bisimulation and a simple boolean modal logic with intersection seems to have the right expressive power to talk about efficiency and stability notions, since all of them are now invariant under ∩-Bisimulations. We now check if the properties define classes of frames that are closed under the different operations introduced. The results can be read off the tables as in the previous section.

Closure Results for class of frames defined by global properties

	BMI	GSF	DU	refl.GSF	BisSysIm		BMI	GSF	DU	refl.GSF	BisSysIm		BMI	GSF	DU	refl.GSF	BisSysIm
$PowG1$	Y	Y	Y	Y	Y	$PrefG4$	Y	Y	N	N	Y	$PPG1$	Y	Y	Y	Y	Y
$PowG2$	Y	Y	Y	Y	Y	$PrefG5$	N	Y	Y	Y	Y	$PPG2$	Y	Y	N	N	Y
$PowG3$	Y	Y	Y	Y	Y	$PrefG6$	N	Y	N	N	Y	$PPG3$	N	N	N	Y	?
$PowG4$	Y	Y	Y	Y	Y	$PrefG7$	N	Y	N	N	Y	$PPG4$	N	Y	Y	Y	Y
$PowG5$	Y	Y	Y	Y	Y	$PrefG8$	Y	Y	Y	Y	Y	$PPG5$	Y	Y	Y	Y	Y
$PrefG1$	Y	Y	Y	Y	Y	$PrefG9$	N	Y	Y	Y	Y	$PPG6$	Y	Y	Y	Y	Y
$PrefG2$	Y	Y	Y	Y	Y	$PrefG10$	Y	Y	Y	Y	Y	$PPG7$	Y	N	Y	Y	Y
$PrefG3$	Y	Y	N	N	Y	$PrefG11$	N(6)	Y	Y	Y	Y						

Proposition 6. *Validity of $PrefG11$ is not preserved under* BMI.

Proof. Consider the frames $\mathcal{F} = \langle \{x, v, w\}, \{1, 2\}, \{ \overset{C}{\to} \mid C \subseteq \{1, 2\}\}, \{\leq_1, \leq_2\}\rangle$, with $w \overset{1}{\to} v, w \overset{2}{\to} v, \overset{\{1,2\}}{\to} = \emptyset, w \leq_1 v, w \leq_2 x$ and $\mathcal{F}' = \langle \{s, t\}, \{1, 2\}, \{ \overset{C}{\to} \mid C \subseteq \{1, 2\}\}, \{\leq_1', \leq_2'\}\rangle$, with $s \overset{1}{\to} t, s \overset{2}{\to} t, \overset{\{1,2\}}{\to} = \emptyset, s \leq_1 t, s \leq_2 t$. Then $f : W \to W', f(w) = s, f(v) = f(x) = t$ is a surjective BM. However, $\mathcal{F} \Vdash PrefG11$ and $\mathcal{F}' \not\Vdash PrefG11$ because $s \leq_1 t, s \leq_2$ and it is not the case that $s \overset{\{1,2\}}{\to} t$.

At the frame level, ML is a fragment of Monadic Second Order Logic. That it does better at this level is thus not only an artifact of the chosen notions.

6 Modal Definability

The previous model-theoretic results give us information about possible definability results. However, let us be more constructive and give formulas that indeed do the job: be it for local-satisfaction or frame-definability aims. Correspondence proofs can be found in the technical report. We indicate the least expressive language we found still being able to express the property under consideration. Another useful criterion is that of the computational complexity of the logic, i.e. of its satisfiability problem (SAT) and model checking problem (MC). Since we lack the space to discuss these issues in depth here is how we bridge our expressivity and complexity results: for each local (resp. global) notion, find the least expressive logic that is still able to express it locally (resp. define the class of frames corresponding to it) and take the complexity of this logic as an *upper bound*. Due to space restrictions we only indicate these upper bounds and references for them. [17] has more details. We assume the reader to be familiar with P, PSPACE and EXPTIME [18]. Π_1^0-complete problems [19] are undecidable but co-recursively enumerable (e.g. $\mathbb{N} \times \mathbb{N}$ tiling [20]).

6.1 Defining Local Notions

Representative definability results

	Axiom	Best Language	SAT	MC
$PowL1$	$\langle C\rangle p$	$\mathcal{L}(N)$	PSPACE[21]	P[22]
$PowL2$	$\bigwedge_{C\not\supseteq i}[C]\neg p$	$\mathcal{L}(N)$	PSPACE[23]	P[22]
$PowL3$	$\downarrow x.[D]\downarrow y.@_x\langle C\rangle y$ (7)	$\mathcal{L}(N,\downarrow,@,x)$	(8) EXPTIME[24]	PSPACE [25]
$PrefL1$	$\langle\leq_i\rangle p$	$\mathcal{L}(N)$	PSPACE[21]	P[22]
$PrefL2$	$\downarrow x.\langle\leq_i\rangle(p \wedge [\leq_i]\neg x)$	$\mathcal{L}(N,\downarrow,x)$	EXPTIME[24]	PSPACE[25]
$PrefL3$	$\downarrow x.\langle\bigcap_{i\in N}\leq_i\rangle(\bigvee_{j\in N}[\leq_j]\neg x)$	$\mathcal{L}(N,\downarrow,\cap,x)$	$\Pi^0_1[11]$	PSPACE
$PrefL4$	$\langle\leq_i^{-1}\rangle p$	$\mathcal{L}(N,\downarrow,@,x)$	PSPACE	PSPACE [25]
$PrefL5$	$\downarrow x.\langle\leq_i^{-1}\rangle(p \wedge [\leq_i^{-1}]\neg x)$	$\mathcal{L}(N,\downarrow,^{-1},x)$	$\Pi^0_1[26]$	PSPACE
$PrefL6$	$[(\leq_i \cap\leq_j) \cup (\leq_j \cap\leq_i)]\bot$	$\mathcal{L}(N,-,\cap)$	EXPTIME[27]	P[28]
$PrefL7$	$\langle\leq_i \cap(\bigcap_{j\in N-\{i\}} \leq_j)\rangle\top$	$\mathcal{L}(N,-,\cap)$	EXPTIME[27]	P[28]
$PrefL8$	$[]\ \leq_i\ []p$	$\mathcal{L}(N,[]\ [])$	EXPTIME[27]	P[28]
$PrefL9$	$\downarrow x.A\downarrow y.(\neg\langle\leq_i\rangle x \wedge @_x\langle\leq_i\rangle y)$	$\mathcal{L}(\downarrow,@,x,E)$	$\Pi^0_1[26]$	PSPACE[29]
$PrefL10$	$[\leq_i]p$	$\mathcal{L}(N)$	PSPACE[23]	P[22]
$PrefL11$	$\downarrow x.[\leq_i]([\leq_i]\neg x \Rightarrow p)$	$\mathcal{L}(\downarrow,x)$	EXPTIME[24]	PSPACE[25]
$PPL1$	$\langle C\cap \leq_i\rangle\top$	$\mathcal{L}(N,\cap)$	PSPACE [30]	P[28]
$PPL2$	$\langle C \cap (\bigcap_{i\in D} \leq_i)\rangle\top$	$\mathcal{L}(N,\cap)$	PSPACE[30]	P[28]
$PPL3$	$\langle(\bigcap_{i\in N}\leq_i) \cap (\bigcup_{C\subseteq N}\xrightarrow{C})\rangle\top$	$\mathcal{L}(N,-,\cap)$	EXPTIME[27]	P[28]
$PPL4$	$[\overline{C}\cap \leq_i]\bot$	$\mathcal{L}(N,-,\cap)$	EXPTIME[27]	P[28]
$PPL5$	$\downarrow x.[\overline{C}\cap \leq_i]\langle\leq_i\rangle x$	$\mathcal{L}(N\downarrow,-,\cap,x)$	$\Pi^0_1[26]$	PSPACE
$PPL6$	$\bigvee_{C\subseteq N}[C \cap \overline{\leq_i}]\bot$	$\mathcal{L}(N,-,\cap)$	EXPTIME[27]	P[28]
$PPL7$	$\downarrow x.[C]\downarrow y.(\neg\langle\leq_i\rangle x \wedge @_x\langle\leq_i\rangle y)$	$\mathcal{L}(N,\downarrow,@,x)$	$\Pi^0_1[24]$	PSPACE[25]
$EF1$	$\downarrow x.[\bigcap_{i\in N}\leq_i]\bigvee_{i\in N}\langle\leq_i\rangle x$	$\mathcal{L}(N,\downarrow,\cap)$	$\Pi^0_1[11,24]$	PSPACE
$EF2$	$\neg\downarrow x.\langle\bigcap_{i\in N}\leq_i\rangle(\bigvee_{j\in N}[\leq_j]\neg x)$	$\mathcal{L}(N,\downarrow,\cap)$	$\Pi^0_1[11,24]$	PSPACE
$EF3$	$[\bigcap_{i\in N}\leq_i]\bot$	$\mathcal{L}(N,\cap)$	PSPACE [30]	P[28]
$ST1$	$\bigwedge_{i\in N}\downarrow x.[i\cap\leq_i]\langle\leq_i\rangle x$	$\mathcal{L}(N,\downarrow,\cap)$	$\Pi^0_1[11]$	PSPACE
$ST2$	$\bigwedge_{i\in N}[i\cap \leq_i]\bot$	$\mathcal{L}(N,\cap)$	PSPACE [30]	P[28]
$ST3$	$\bigwedge_{C\subseteq N}\downarrow x.[C \cap (\bigcap_{i\in C} \leq_i)]\bigvee_{j\in C}\langle\leq_j\rangle x$	$\mathcal{L}(N,\downarrow,\cap)$	$\Pi^0_1[24]$	PSPACE
$ST4$	$\bigwedge_{C\subseteq N}[C \cap (\bigcap_{i\in C} \leq_i)]\bot$	$\mathcal{L}(N,\cap)$	PSPACE [30]	P[28]

Proposition 7. *$PowL3$ is true of $\mathcal{M},w$ iff $\mathcal{M},w,g \Vdash\downarrow x.[D] \downarrow y.@_x\langle C\rangle y$.*

Proof. From right to left: Assume that $\mathcal{M},w,g \Vdash\downarrow x.[D] \downarrow y.@_x\langle C\rangle y$. Then $\mathcal{M},w,g[x := w], \Vdash [D] \downarrow y.@_x\langle C\rangle y$. But now assume there is a state v that coalition D can force from w. By definition, $w \xrightarrow{D} v$ (1). But by (1) and semantics of $[D]$ then we have $\mathcal{M},v,g[x := w], \Vdash\downarrow y.@_x\langle C\rangle y$ (2). (2) and semantics of $\downarrow$ gives us $\mathcal{M},v,g[x := w, y := v] \Vdash @_x\langle C\rangle y$ (3). From (3) and semantics of $@_x$ and the fact that $g(x) = w$ we have $\mathcal{M},w,g[x := w, y := v] \Vdash \langle C\rangle y$ (4). But by semantics of $\langle C\rangle$ and the fact that $g(y) = v$, (4) really means that $w \xrightarrow{C} v$ (5). Since the v was arbitrary, it follows from (5) that at w for any state v, if D can achieve it, then C can do so, too. But this precisely means that $PowL3$ is true of $\mathcal{M},w$. □

Theorem 1 (ten Cate [24]). *The satisfiability problems for formulas in $\mathcal{ML}(N,\downarrow,@,x) - \Box \downarrow\Box$ with bounded width is EXPTIME-complete.*

Proposition 8. *PowL3 is expressible in an extended modal language with a satisfiability problem in* EXPTIME.

Proof. By the previous proposition, we have *PowL3* is defined by $\downarrow x.[\text{D}] \downarrow y.@_x \langle \text{C} \rangle y$. But $\downarrow x.[\text{D}] \downarrow y.@_x \langle \text{C} \rangle y$ does contain the $\Box \downarrow \Box$ scheme. Thus, *PowL3* is defined by a formula in $\mathcal{ML}(\text{N}, \downarrow, @, x) - \Box \downarrow \Box$ (1). But by Theorem 1 the satisfiability problem of $\mathcal{ML}(\text{N}, \downarrow, @, x) - \Box \downarrow \Box$ is in EXPTIME.   □

6.2 Defining Global Notions

First of all, we define what it means for a formula to be valid on a class of frames.

Definition 12 (Validity on a class of frames). *We say that a formula ϕ is valid on a class of frames* F *iff for any frame $\mathcal{F} \in$ F *and any model $\mathcal{M}$ based on $\mathcal{F}$, at all states w in $Dom(\mathcal{F})$, $\mathcal{M}, w \Vdash \phi$. We write* F $\Vdash \phi$.

Modal definability has again two sides: We can look for a formula ϕ such that $\mathcal{M}, w \Vdash \phi$ iff $\mathcal{M}, w$ has some property, or such that F $\Vdash \phi$ iff F has the property.

	Axiom	Best Language	SAT	MC
$PowG1$	$\bigwedge_{\text{C} \subseteq \text{N}}((\langle \text{C} \rangle \phi \Rightarrow [\text{C}]\phi) \wedge \langle \text{C} \rangle \top)$	$\mathcal{L}(\text{N})$	PSPACE[21]	P[22]
$PowG2$	$\bigwedge_{\text{C}:|\text{C}|<|\text{N}|/2}[\text{C}]\bot$	$\mathcal{L}(\text{N})$	PSPACE[23]	P[22]
$PowG3$	$\bigwedge_{\text{C} \subseteq \text{N}}(\langle \text{C} \rangle \phi \Rightarrow [\text{C}]\phi)$	$\mathcal{L}(\text{N})$	PSPACE[21]	P[22]
$PowG4$	$\bigwedge_{\text{C} \subseteq \text{N}} \bigwedge_{\text{D} \supseteq \text{C}}(\langle \text{C} \rangle \phi \Rightarrow \langle \text{D} \rangle \phi)$	$\mathcal{L}(\text{N})$	PSPACE[23]	P[22]
$PowG5$	$\langle \text{C} \rangle \top \Rightarrow \bigwedge_{\text{D}:\text{C} \cap \text{D}=\emptyset}((\langle \text{D} \rangle \phi \Rightarrow \langle \text{C} \rangle \phi)$	$\mathcal{L}(\text{N})$	PSPACE[21]	P[22]
$PrefG1$	$\phi \Rightarrow \langle \leq_i \rangle \phi$	$\mathcal{L}(\text{N})$	PSPACE[23]	P[22]
$PrefG2$	$\langle \leq_i \rangle \langle \leq_i \rangle \phi \Rightarrow \langle \leq_i \rangle \phi$	$\mathcal{L}(\text{N})$	PSPACE[21]	P[22]
$PrefG3$	$(p \wedge \text{E}q) \Rightarrow (\text{E}(p \wedge \langle \leq_i \rangle q) \vee \text{E}(q \wedge \langle \leq_i \rangle p))$	$\mathcal{L}(\text{N}, \text{E})$	EXPTIME[31]	P[28]
$PrefG4$	Conjunction of the 3 previous axioms	$\mathcal{L}(\text{N}, \text{E})$	EXPTIME[32]	P[28]
$PrefG5$	see below	$\mathcal{L}(\text{N})$	PSPACE[23]	P[22]
$PrefG6$	$\bigwedge_{i \in \text{N}}(@_j \langle \leq_i \rangle k \vee @_k j \vee @_k \langle \leq_i \rangle j)$	$\mathcal{L}(\text{N}, @, i)$	PSPACE[33]	P[25]
$PrefG7$	$PrefG5 \wedge PrefG6 \wedge (\bigwedge_{i \in \text{N}}(j \Rightarrow \neg \langle \leq_i \rangle j))$	$\mathcal{L}(\text{N}, @, i)$	PSPACE[33]	P[25]
$PrefG8$	$\bigwedge_{i \in \text{N}}((\langle \leq_i \rangle \phi \Rightarrow [\leq_i]\phi) \wedge \langle \leq_i \rangle \top)$	$\mathcal{L}(\text{N})$	PSPACE[23]	P[22]
$PrefG9$	$\langle \text{C} \rangle j \leftrightarrow \bigwedge_{i \in \text{C}} \langle \leq_i \rangle j$	$\mathcal{L}(\text{N}, i)$	PSPACE[33]	P[25]
$PrefG10$	$\langle \text{C} \rangle p \leftrightarrow \bigvee_{i \in \text{C}} \langle \leq_i \rangle p$	$\mathcal{L}(\text{N})$	PSPACE[21]	P[22]
$PrefG11$	$\langle \text{C} \rangle j \leftrightarrow \bigvee_{\text{D} \subseteq \text{C} \& |\text{D}| > \frac{|\text{C}|}{2}}(\bigwedge_{i \in \text{D}} \langle \leq_i \rangle j)$	$\mathcal{L}(\text{N}, i)$	PSPACE[33]	P[25]
$PPG1$	$\langle \text{C} \rangle \phi \Rightarrow \bigwedge_{i \in \text{N}} \langle \leq_i \rangle \phi$	$\mathcal{L}(\text{N})$	PSPACE[23]	P[22]
$PPG2$	$\bigvee_{i \in \text{N}} \text{A} \bigwedge_{\text{C} \subseteq \text{N}}(\langle \text{C} \rangle \phi \Rightarrow \langle \leq_i \rangle \phi)$	$\mathcal{L}(\text{N}, \text{E})$	EXPTIME[31]	P[28]
$PPG3$	$\bigwedge_{i \in \text{N}} \bigvee_{\text{C} \subseteq \text{N}} (\overline{\leq_i} \cup \leq_i) \langle \overline{\leq_i} \cap \text{C} \rangle \top$	$\mathcal{L}(-, \cap, \cup)$	EXPTIME[27]	P[28]
$PPG4$	see below	$\mathcal{L}(\text{N}, i)$	PSPACE[33]	P[25]
$PPG5$	$\bigwedge_{\text{C} \not\supseteq \{i\}} [\overset{\text{C}}{\rightarrow}]\bot$	$\mathcal{L}(\text{N})$	PSPACE[21]	P[22]
$PPG6$	$\langle \text{C} \rangle \phi \Rightarrow \bigvee_{\text{D} \subset \text{C}} \langle \text{D} \rangle \phi$	$\mathcal{L}(\text{N})$	PSPACE[23]	P[22]
$PPG7$	$\bigwedge_{i \in \text{N}} \text{E} \bigwedge_{\text{C} \not\supseteq \{i\}} [\text{C}]\bot$	$\mathcal{L}(\text{N}, \text{E})$	EXPTIME[33]	P[25]

$$\bigwedge_{i \in \text{N}} (j \wedge \langle \leq_i \rangle (k \wedge \neg \langle \leq_i \rangle j \wedge \langle \leq_i \rangle (l \wedge \neg \langle \leq_i \rangle k))) \Rightarrow j \wedge \langle \leq_i \rangle (l \wedge \neg \langle \leq_i \rangle j) \quad (AxPrefG5)$$

$$[p \wedge \langle i \rangle q \wedge \langle \leq_i \rangle (q \wedge \langle \leq_i \rangle \neg p)] \Rightarrow \bigwedge_{i \in \text{C} \subseteq \text{N}} [(\langle \text{C} \rangle r \wedge \bigwedge_{\text{D} \subseteq \text{C} \setminus i} \neg \langle \text{D} \rangle r) \Rightarrow \langle \leq_i \rangle (r \wedge \neg \langle \leq_i \rangle p)] \quad (PPG4)$$

7   Conclusion

We identified a set of natural notions for reasoning about cooperation: local notions giving properties of a state of a given system and global notions defining a class of frames. We provided satisfiability (resp. validity) invariance results for these notions for a large class of operations and relations between models (resp. frames). We also gave explicit definability results and observed that defining frames for cooperation logics does not seem too demanding in terms of expressive power, as most of the notions considered are definable in the basic modal language. On the other hand, our results show that local notions call for modal logics for which satisfaction is not invariant under bounded morphisms. However, as long as we avoid converse modalities, interesting reasoning about cooperation can be done within GSM-invariant modal languages. Though this fact does not directly lead to a nice upper bound on the complexity of the logic's SAT (nor to its decidability), our definability results show that most of the considered notions can (individually) be expressed in MLs in EXPTIME. Moreover, for several notions we only found logics with undecidable SAT that could express them. All these notions involve the idea of a "strict" improvement (e.g. Nash-stable, Pareto-efficient). By contrast, strong notions of stability and efficiency (EF3, ST2, ST4) are all expressible in logics with SAT in PSPACE. Thus, we could say that "expressing strictness" and therefore "weak" notions are dangerous, while "strong" notions (looking only at the weak preference relation) are safe.

   Based on our current work, the following lines seem worth exploring:

- Since dealing with real coalitional powers is probably more natural using neighborhood semantics, it will be useful to do the same work for modal logics of the **CL**-type or of the type of one of its normal simulations [2].
- It would be interesting to obtain similar invariance results and upper bounds on the complexity of the logics needed to encode *concrete arguments* from SCT and (cooperative) GT, thus addressing the complexity of *actual reasoning* about cooperative situations.
- From our definability results we could obtain upper bounds (on SAT) and on the *combined complexity* of model checking of logics able to express certain notions from SCT and GT. The converse road would be to use complexity results from computational social choice and algorithmic game theory to obtain lower bounds on its *data complexity*.[4] As an example: a way to go could be to take a hardness result for the problem of determining whether a profile of strategies is a pure Nash-equilibrium of given game (with respect to some reasonable and qualitative encoding of games) as a lower bound on the data complexity of model-checking of a logic than can express this notion.
- In order to obtain a complete picture of the complexity of reasoning about cooperation, we need a procedure to assess the LB of the complexity of modal logics that can express some notion.

---

[4] When measuring combined complexity both the formula and the model are part of the input. While when measuring data complexity, the formula is fixed and the model is the input (see [34]).

- Our definability results made use of very big conjunctions and disjunctions. It would be interesting to check how the length of these formulas is related to a more reasonable input such as the number of agents. (When taking conjunction/disjunctions over all coalitions, they will be exponentially related.)
- We could also consider the complexity effects of using more succinct languages that have more modalities, e.g. a modality $\langle Most \leq \rangle \phi$, read: "there is a ϕ-state that a majority of agents finds at least as good as the current one" (cf. e.g. [35]).

Acknowledgments. The authors are very grateful to Johan van Benthem, Natasha Alechina, Jakub Szymanik, Eric Pacuit, Peter van Emde Boas, Balder ten Cate, Ulle Endriss and the anonymous referees for their useful comments.

References

1. Pauly, M.: A modal logic for coalitional power in games. JLC 12(1), 149–166 (2002)
2. Broersen, J., Herzig, A., Troquard, N.: Normal Coalition Logic and its conformant extension. In: Samet, D. (ed.) TARK 2007, PUL, pp. 91–101 (2007)
3. Girard, P.: Modal Logic for Preference Change. PhD thesis, Stanford (2008)
4. Kurzen, L.: Logics for Cooperation, Actions and Preferences. Master's thesis, Universiteit van Amsterdam, The Netherlands (2007)
5. Ågotnes, T., Dunne, P.E., van der Hoek, W., Wooldridge, M.: Logics for coalitional games. In: LORI 2007, Beijing, China (2007) (to appear)
6. Alur, R., Henzinger, T., Kupferman, O.: Alternating-time temporal logic. In: Proceedings of the 38th IEEE Symposium on Foundations of Computer Science, Florida (October 1997)
7. Goranko, V.: Coalition games and alternating temporal logics. In: TARK 2001, pp. 259–272. Morgan Kaufmann, San Francisco (2001)
8. Hansen, H.H., Kupke, C., Pacuit, E.: Bisimulation for neighbourhood structures. In: Mossakowski, T., Montanari, U., Haveraaen, M. (eds.) CALCO 2007. LNCS, vol. 4624, pp. 279–293. Springer, Heidelberg (2007)
9. de Jongh, D., Liu, F.: Optimality, belief and preference. In: Artemov, S., Parikh, R. (eds.) Proc. of the Workshop on Rationality and Knowledge, ESSLLI (2006)
10. Osborne, M.J., Rubinstein, A.: A course in game theory. MIT Press, Cambridge (1994)
11. Cate, B.: Model theory for extended modal languages. PhD thesis, University of Amsterdam, ILLC Dissertation Series DS-2005-01 (2005)
12. Blackburn, P., de Rijke, M., Venema, Y.: Modal Logic. Cambridge Tracts in Theoretical Computer Science, vol. 53. Cambridge University Press, UK (2001)
13. van Benthem, J.: Modal Logic and Classical Logic. Bibliopolis, Napoli (1983)
14. Areces, C., Blackburn, P., Marx, M.: Hybrid logics: characterization, interpolation and complexity. The Journal of Symbolic Logic 66(3), 977–1010 (2001)
15. Feferman, S.: Persistent and invariant formulas for outer extensions. Compositio Mathematica 20, 29–52 (1969)
16. Goldblatt, R.I., Thomason, S.K.: Axiomatic classes in propositional modal logic. In: Crossley, J.N. (ed.) Algebra and Logic: Papers 14th Summer Research Inst. of the Australian Math. Soc. LNM, vol. 450, pp. 163–173. Springer, Berlin (1975)

17. Dégremont, C., Kurzen, L.: Modal logics for preferences and cooperation: Expressivity and complexity. ILLC PP-2008-39, University of Amsterdam (2008), http://staff.science.uva.nl/~cdegremo/

18. Papadimitriou, C.M.: Computational complexity. Addison-Wesley, MA (1994)

19. Odifreddi, P.: Classical Recursion Theory. Studies in Logic and the Foundations of Mathematics, vol. (125). North-Holland, Amsterdam (1989)

20. Harel, D.: Recurring dominoes: making the highly undecidable highly understandable. In: Topics in the theory of computation, New York, NY, USA, pp. 51–71. Elsevier, Amsterdam (1985)

21. Ladner, R.E.: The computational complexity of provability in systems of modal propositional logic. SIAM J. Comput. 6(3), 467–480 (1977)

22. Fischer, M.J., Ladner, R.E.: Propositional dynamic logic of regular programs. J. Comput. Syst. Sci. (1979)

23. Halpern, J.Y., Moses, Y.: A guide to completeness and complexity for modal logics of knowledge and belief. Artificial Intelligence 54(3), 319–379 (1992)

24. ten Cate, B., Franceschet, M.: On the complexity of hybrid logics with binders. In: Ong, L. (ed.) CSL 2005. LNCS, vol. 3634, pp. 339–354. Springer, Heidelberg (2005)

25. Franceschet, M., de Rijke, M.: Model checking for hybrid logics. In: Proceedings of the Workshop Methods for Modalities (2003)

26. Gurevich, Y.: The classical decision problem. In: Perspectives in Mathematical Logic. Springer, Heidelberg (1997)

27. Lutz, C., Sattler, U.: The complexity of reasoning with boolean modal logics. In: Wolter, W., de Rijke, Z. (eds.) AiML, WS, pp. 329–348 (2000)

28. Lange, M.: Model checking pdl with all extras. J. Applied Logic 4(1), 39–49 (2006)

29. Franceschet, M., de Rijke, M.: Model checking hybrid logics (with an application to semistructured data). J. Applied Logic 4(3), 279–304 (2006)

30. Donini, F.M., Lenzerini, M., Nardi, D., Nutt, W.: The complexity of concept languages. In: KR, pp. 151–162 (1991)

31. Spaan, E.: Complexity of modal logics. PhD thesis, ILLC Amsterdam (1993)

32. Hemaspaandra, E.: The price of universality. NDJFL 37(2), 174–203 (1996)

33. Areces, C., Blackburn, P., Marx, M.: A road-map on complexity for hybrid logics. In: Flum, J., Rodríguez-Artalejo, M. (eds.) CSL 1999. LNCS, vol. 1683, pp. 307–321. Springer, Heidelberg (1999)

34. Vardi, M.Y.: The complexity of relational query languages (extended abstract). In: STOC 1982: Proceedings of the fourteenth annual ACM symposium on Theory of computing, pp. 137–146. ACM, New York (1982)

35. Ågotnes, T., van der Hoek, W., Wooldridge, M.: Quantified coalition logic. In: Veloso, M.M. (ed.) Proceedings of the Twentieth International Joint Conference on Artificial Intelligence (IJCAI 2007), California, pp. 1181–1186. AAAI Press, Menlo Park (2007)

Simulation and Information: Quantifying over Epistemic Events

Hans van Ditmarsch[1,*] and Tim French[2]

[1] Philosophy, University of Sevilla, Spain
hvd@us.es
[2] Computer Science and Software Engineering,
University of Western Australia, Australia
tim@csse.uwa.edu.au

Abstract. We introduce a multi-agent logic of knowledge with time where $F\varphi$ stands for "there is an informative event after which φ." Formula $F\varphi$ is true in a model iff it is true in all its refinements (i.e., *atoms* and *back* are satisfied; the dual of simulation). The logic is *almost* normal, and positive knowledge is preserved. The meaning of $F\varphi$ is also "after the agents become aware of new factual information, φ is true," and on finite models it is also "there is an event model (M, s) after which φ." The former provides a correspondence with bisimulation quantifiers in a setting with epistemic operators.

Keywords: Bisimulation quantifier, modal logic, temporal epistemic logic, multi-agent system.

1 Introduction

If you know where you are and you know what's going to happen, you *want* to know where you will end up. But it can also be that you know where you are and know where you would like to end up, and that you *want* to know how to make that happen. Or you might *want* to know where you can end up in the first place, disregarding how that may be brought about. In the setting of logics for information update [1,2,3], knowledge of where you are and where you end up is formalized in multi-agent epistemic logic and semantically represented by a pointed multi-agent Kripke model, and knowledge about what's going to happen is formalized as a dynamic modal operation that is interpreted as a relation between such Kripke models. The standard focus in dynamic epistemic logic was on the first of the three issues above: precision about a specific information update and precision about the effects of that update. In this contribution we

* The authors gratefully acknowledge the suggestions and insights from the anonymous LOFT 2008 referees and from the anonymous KRAMAS 2008 referees. Hans van Ditmarsch acknowledges support of the Netherlands Institute of Advanced Study where he was Lorentz Fellow in 2008. Hans van Ditmarsch is also affiliated to the University of Otago. Part of this work was carried out while affiliated to CNRS-IRIT, Université de Toulouse, France.

J.-J.Ch. Meyer and J.M. Broersen (Eds.): KRAMAS 2008, LNAI 5605, pp. 51–65, 2009.

focus on the other two issues instead. As this is partly about what may happen after *any* event, this concerns quantification over events. Our work is a further generalization of works such as [4],[5] (and its final journal version [6])and our presentation of future event operators as temporal is motivated by works such as [7] linking temporal epistemic logic to dynamic epistemic logic.

We introduce a very succinct logic of future events: the multi-agent logic of knowledge with (only) an operation $G\varphi$ that stands for "φ holds after all informative events" — the diamond version $F\varphi$ stands for "there is an informative event after which φ."

Previous works [8,9] have modelled informative events using a notion of model refinement. In [9] it was shown that model restrictions were not sufficient to simulate informative events, and they introduced *refinement trees* for this purpose—a precursor of the semantics of dynamic epistemic logics developed later [2]. We incorporate implicit quantification over informative events directly into the language using a similar notion of *refinement*; in our case a refinement is the inverse of simulation [10].

We demonstrate that this is useful notion for informative event by a number of technical results for this logic—the logic is *almost* normal, positive knowledge is preserved—and by a number of equivalence results for alternative semantics: $F\varphi$ also means "there is an event model (M, s) after which φ," and it also means "after the agents become aware of new factual information, φ is true." The last provides a correspondence with bisimulation quantifiers [11,12] in a setting with epistemic operators, as in [13].

2   Technical Preliminaries

Structural notions. Assume a finite set of agents A and a countably infinite set of atoms P.

Definition 1 (Structures). *An epistemic model $M = (S, R, V)$ consists of a domain S of (factual) states (or worlds), accessibility $R : A \to \mathcal{P}(S \times S)$, and a valuation $V : P \to \mathcal{P}(S)$. For $s \in S$, (M, s) is an epistemic state (also known as a pointed Kripke model).*

For $R(a)$ we write R_a; accessibility R can be seen as a set of relations R_a, and V as a set of valuations $V(p)$. Given two states s, s' in the domain, $R_a(s, s')$ means that in state s agent a considers s' a possibility. We adopt the standard rules for omission of parentheses in formulas, and we also delete them in representations of structures such as (M, s) whenever convenient and unambiguous.

Definition 2 (Bisimulation, simulation, refinement). *Let two models $M = (S, R, V)$ and $M' = (S', R', V')$ be given. A non-empty relation $\mathfrak{R} \subseteq S \times S'$ is a bisimulation, iff for all $s \in S$ and $s' \in S'$ with $(s, s') \in \mathfrak{R}$:*

atoms $s \in V(p)$ *iff* $s' \in V'(p)$ *for all* $p \in P$
forth *for all* $a \in A$ *and all* $t \in S$, *if* $R_a(s, t)$, *then there is a* $t' \in S'$ *such that* $R_a(s', t')$ *and* $(t, t') \in \mathfrak{R}$

back *for all $a \in A$ and all $t' \in S'$, if $R_a(s', t')$, then there is a $t \in S$ such that $R_a(s, t)$ and $(t, t') \in \mathfrak{R}$*

We write $(M, s) \leftrightharpoons (M', s')$, iff there is a bisimulation between M and M' linking s and s'. Then we call (M, s) and (M', s') bisimilar. We also say that (M, s) is similar to (M', s') and vice versa.

A relation that satisfies **atoms** *and* **forth** *is a simulation, and in that case (M', s') is a simulation of (M, s), and (M, s) is a refinement of (M', s), and we write $(M, s) \rightharpoonup (M', s')$ (or $(M', s') \leftharpoondown (M, s)$).*

A bisimulation (simulation) that satisfies **atoms** *for a subset $P' \subseteq P$ is a P'-bisimulation (P'-simulation); we write $(M, s) \leftrightharpoons_{P'} (M', s')$ $((M, s) \rightharpoonup_{P'} (M', s'))$, etc.*

Standard language notions. The languages of propositional logic ($\mathcal{L}_{pl}$) and of epistemic logic ($\mathcal{L}_{el}$) — $a \in A$, $p \in P$, $B \subseteq A$.

$$\mathcal{L}_{pl} : \qquad\qquad \varphi ::= p \mid \neg\varphi \mid (\varphi \wedge \varphi)$$
$$\mathcal{L}_{el} : \qquad\qquad \varphi ::= p \mid \neg\varphi \mid (\varphi \wedge \varphi) \mid K_a\varphi$$

Standard abbreviations include: $\varphi \vee \psi$ iff $\neg(\neg\varphi \wedge \neg\psi)$; $\varphi \to \psi$ iff $\neg\varphi \vee \psi$, $\hat{K}_a\varphi$ iff $\neg K_a \neg\varphi$.

Event model logic. All the following are simultaneously defined:

Definition 3. *Language $\mathcal{L}_{eml}$ of event model logic:*

$$\varphi ::= p \mid \neg\varphi \mid \varphi \wedge \varphi \mid K_a\varphi \mid [\mathsf{M}, \mathsf{s}]\varphi$$

Definition 4 (Event model). *An event model for a finite set of agents A and a language $\mathcal{L}$ is a triple $\mathsf{M} = (\mathsf{S}, \mathsf{R}, \mathsf{pre})$ where*

- *domain S is a finite non-empty set of events,*
- *$\mathsf{R} : A \to \mathcal{P}(\mathsf{S} \times \mathsf{S})$ assigns an accessibility relation to each agent,*
- *$\mathsf{pre} : \mathsf{S} \to \mathcal{L}$ assigns to each event a precondition,*

A pair (M, s) with a distinguished actual event $\mathsf{s} \in \mathsf{S}$ is called an epistemic event. An epistemic event with a singleton domain, accessible to all agents, and identity postcondition, is a public announcement.

Definition 5 (Semantics of event model logic). *Let a model (M, s) with $M = (S, R, V)$ be given. Let $a \in A$, $B \subseteq A$, and $\varphi, \psi \in \mathcal{L}$.*

$$(M, s) \models [\mathsf{M}, \mathsf{s}]\varphi \text{ iff } (M, s) \models \mathsf{pre}(\mathsf{s}) \text{ implies } (M \otimes \mathsf{M}, (s, \mathsf{s})) \models \varphi$$

Definition 6 (Execution of an event model). *Given are an epistemic model $M = (S, R, V)$, a state $s \in S$, an event model $\mathsf{M} = (\mathsf{S}, \mathsf{R}, \mathsf{pre})$, and an event $\mathsf{s} \in \mathsf{S}$ with $(M, s) \models \mathsf{pre}(\mathsf{s})$. The result of executing (M, s) in (M, s) is the model $(M \otimes \mathsf{M}, (s, \mathsf{s})) = ((S', R', V'), (s, \mathsf{s}))$ where*

- *$S' = \{(t, \mathsf{t}) \mid (M, t) \models \mathsf{pre}(\mathsf{t})\}$,*
- *$R'(a) = \{((t, \mathsf{t}), (u, \mathsf{u})) \mid (t, \mathsf{t}), (u, \mathsf{u}) \in S' \text{ and } (t, u) \in R(a) \text{ and } (\mathsf{t}, \mathsf{u}) \in \mathsf{R}(a)\}$,*
- *$V'(p) = \{(t, \mathsf{t}) \mid (M, t) \models p\}$.*

Bisimulation quantifiers and bisimulation quantified epistemic logic. The language and semantics are as follows.

Definition 7 (Bisimulation quantified epistemic logic). *Bisimulation quantified epistemic logic augments epistemic logic by additionally allowing formulas of the kind $\forall p\varphi$ in the recursive definition, where p is an atom of P, and φ is a formula. This is the language $\mathcal{L}_{bqel}$.*

Given an epistemic model $M = (S, R, V)$ and a state $s \in S$ we say:

$$M, s \models \forall p\varphi \quad \textit{iff} \quad \textit{for every epistemic model } (M', s') \leftrightarrows_{P\setminus\{p\}} (M, s) : M', s' \models \varphi.$$

3   Simulation and Information

In *future event logic* one can express what informative events can be expected in a given information state. The language and the semantics of future event logic are as follows.

Definition 8 (Language $\mathcal{L}_{fel}$). *Given agents A and atoms P, the language $\mathcal{L}_{fel}$ is inductively defined as*

$$\varphi ::= p \mid \neg\varphi \mid (\varphi \wedge \varphi) \mid K_a\varphi \mid G\varphi$$

where $a \in A$ and $p \in P$.

We write $F\varphi$ for $\neg G\neg\varphi$. We propose a dynamic epistemic modal way to interpret temporal operators. This means that our future is the *computable future*: $F\varphi$ is true now, iff there is an (unspecified) informative event after which φ is true.

In the semantics for $G\varphi$, now to follow, we use the structural notion of *refinement*, see Definition 2. From the requirements for bisimulation, a refinement satisfies **atoms** and **back**. Refinement is the dual of *simulation*, that satisfies **atoms** and **forth**. If (M', s') is a refinement of (M, s) we write $(M, s) \leftharpoonup (M', s')$.

Definition 9 (Semantics of future event logic). *Assume an epistemic model $M = (S, R, V)$. The interpretation of $\varphi \in \mathcal{L}_{fel}$ is defined by induction.*

$$
\begin{aligned}
&M, s \models p &&\textit{iff } s \in V_p \\
&M, s \models \neg\varphi &&\textit{iff } M, s \not\models \varphi \\
&M, s \models \varphi \wedge \psi &&\textit{iff } M, s \models \varphi \textit{ and } M, s \models \psi \\
&M, s \models K_a\varphi &&\textit{iff for all } t \in S : (s, t) \in R_a \textit{ implies } M, t \models \varphi \\
&M, s \models G\varphi &&\textit{iff for all } (M', s') : (M, s) \leftharpoonup (M', s') \textit{ implies } M', s' \models \varphi
\end{aligned}
$$

In other words, $G\varphi$ is true in an epistemic state iff φ is true in all of its *refinements*. Note the wrong direction in the definition: the future epistemic state simulates the current epistemic state. Typical model operations that produce a refinement are: blowing up the model (to a bisimilar model) such as adding copies that are indistinguishable from the current model and one another for some agent(s), removing states, and removing pairs of the accessibility relation for an agent. Validity in a model, and validity, are defined as usual. For $\{s \mid M, s \models \varphi\}$ we write $[\![\varphi]\!]_M$.

Example 1. Given are two agents that are uncertain about the value of a fact p, and where this is common knowledge, and where p is true. We assume that both accessibility relations are equivalence relations, and that the epistemic operators model the agents' knowledge. An informative event is possible after which a knows that p but b does not know that: $M, 1 \models F(K_a p \wedge \neg K_b K_a p)$.

In the figure, $(M, 1)$ is the structure on the left, and its refinement validating the postcondition is on the right. In this visualization, the actual state is underlined, states are suggestively named with the value of p (so on the right, the two *different* states where p is true are both named 1), states that are indistinguishable for an agent are linked and labeled with the name of that agent, and transitivity is assumed (so on the right, all three states are indistinguishable for agent b). Note that on the left, the formula $F(K_a p \wedge \neg K_b K_a p)$ is true, because on the right $K_a p \wedge \neg K_b K_a p$ is true: in the underlined actual state there is no alternative for agent a, so $K_a p$ is true, whereas agent b considers it possible that the top-state also named 1 is the actual state, and in that state agent a considers it possible that p is false. Therefore, $\neg K_b K_a p$ is also true in the underlined 1-state.

$$0 - ab - 1$$
$$|$$
$$b$$
$$|$$
$$0 - ab - \underline{1} \quad \Rightarrow \quad \underline{1}$$

Proposition 1. *Some elementary validities are:*

1. $\models G(\varphi \to \psi) \leftrightarrow (G\varphi \to G\psi)$
2. $\models G\varphi \to \varphi$
3. $\models G\varphi \to GG\varphi$
4. $\models \varphi$ *implies* $\models G\varphi$
5. $\models K_a G\varphi \to GK_a\varphi$

Proof.

1. Obvious.
2. A model is a refinement of itself; this corresponds to the trivial event 'announce true'.
3. Consider the diamond version $FF\varphi \to F\varphi$. The relational composition of two simulations is again a simulation.
4. Obvious.
5. Consider the diamond version. Choose an accessible state in a refinement of a model. By **back**, this accessibility step can also be taken in the initial model.

Proposition 1 makes clear that G comes close to being a normal modal operator. But it is not a normal modal logic: the validities of the logic are not closed under uniform substitution of atomic variables for other formulas. For example, given some atom p, $p \to Gp$ is valid, but $\neg Kp \to G\neg Kp$ is not valid. A countermodel

of the latter is the typical two-state situation where there is uncertainty about
the value of p and where p is true. In that case, restriction to the p-state (public
announcement of p) makes it known. Another countermodel is provided by the
example above, for the knowledge of agent b.

A standard check for our bold claim that G formalizes a notion of informative
event is that

Proposition 2. *Bisimilar epistemic states have the same logical theory.*

Proof. This is not completely trivial, because bisimilarity is with respect to
the epistemic operators, whereas the same logical theory is with respect to the
epistemic operators and the temporal operator. Both can be established easily
by the observation that if an epistemic state is a refinement of one of two given
bisimilar epistemic states, it is also a refinement of the other epistemic state,
because the relational composition of a simulation relation and a bisimulation
relation is a simulation relation. The inductive case $G\varphi$ of the proof is:

Assume $\mathfrak{R} : (M, s) \leftrightarrow (M', s')$, and let $M, s \models G\varphi$. To show that $M', s' \models G\varphi$,
let (M'', s'') be such that $\mathfrak{R}' : (M'', s'') \Rightarrow (M', s')$. We now have that $\mathfrak{R}' \circ \mathfrak{R}^{-1} :$
$(M'', s'') \Rightarrow (M, s)$. From that and $M, s \models G\varphi$ follows $M'', s'' \models \varphi$.

The positive formulas are those in the inductively defined fragment

$$\varphi ::= p \,|\, \neg p \,|\, \varphi \vee \varphi \,|\, \varphi \wedge \varphi \,|\, K_a \varphi \,|\, G\varphi.$$

The preserved formulas are those for which

$$\varphi \to G\varphi \text{ is valid.}$$

I.e., they preserve truth under model refinement as long as the refinement includes an image for the actual state; the better known setting is model restriction.
The first real corroboration that the temporal operators formalize a notion of
informative event is that they model *growth of information* in the sense that
positive knowledge does not get lost:

Proposition 3. *Positive formulas preserve truth under refinement of models.*

Proof. Elementary.

Further corroboration that the temporal operators are quantifying over informative events is provided by the observation that *a restricted modal product is a
refinement of a model* if the valuations of the states in that model are preserved
under the product operation. This entails that the execution of an event model
in an epistemic state is a refinement of that epistemic state. This we will now
address.

4   Quantifying over Event Models

An informative update is the execution of an event model in an epistemic state. We consider event models for the epistemic language $\mathcal{L}_{el}$.

Proposition 4. *An informative update is a refinement.*

Proof. Let $(\mathsf{M},\mathsf{s}) = ((\mathsf{S},\mathsf{R},\mathsf{pre}),\mathsf{s})$ be an event model for language $\mathcal{L}_{el}$. Let $(M,s) = ((S,R,V),s)$ be an epistemic state and suppose $M,s \models \mathsf{pre}(\mathsf{s})$. Then $\mathfrak{R}(t,\mathsf{t}) = t$ is a simulation between $((M \otimes \mathsf{M}),(s,\mathsf{s}))$ and (M,s); below we assume that $(M \otimes \mathsf{M}) = (S',R',V')$.

- **atoms:** if $(t,\mathsf{t}) \in V'(p)$ then $t \in V(p)$;
- **forth:** let $((t,\mathsf{t}),(t',\mathsf{t}')) \in R'_a$; then $(t,t') \in R_a$.

Subject to the restrictions that the epistemic models are finite, the fit is exact: refinements are informative updates.

Proposition 5. *(On finite epistemic models) A refinement is an informative update.*

Proof. Given are a finite epistemic state $((S,R,V),s)$ and a refinement $((S', R',V'),s')$ of that model (according to refinement $\mathfrak{R}$). Consider the event model that is isomorphic to that refinement (according to isomorphism $\mathfrak{I}$). Instead of valuations for states t, this event model has preconditions for events $\mathfrak{I}(t)$. We want the preconditions only to be satisfied in states s such that $(s,t) \in \mathfrak{R}$—this we cannot guarantee, but we can come close enough. In a finite model, every state s can be *distinguished* from all other (except bisimilar) states by a formula φ_s. This is shown in Lemma 1 below. These distinguishing formulas can now be used to have event $\mathfrak{I}(t)$ of the event model executable in every state $s \in S$ such that $(s,t) \in \mathfrak{R}$, by requiring that

$$\mathsf{pre}(\mathfrak{I}(t)) = \bigvee_{(s,t)\in\mathfrak{R}} \varphi_s$$

This may give us pairs $(s,\mathfrak{I}(t))$ with $(s,t) \notin \mathfrak{R}$, but in that case s will be bisimilar to some s' satisfying the same distinguishing formula and such that $(s',t) \in \mathfrak{R}$. Of course, the composition of the total bisimulation on (S,R,V) with the refinement relation $\mathfrak{R}$ will also be a refinement relation. Without loss of generality we assume that $\mathfrak{R}$ is maximal in the sense that it is a fixed-point of composition with that total bisimulation. This makes the structure of the proof clearer.

We now show that the restricted modal product $((S'',R'',V''),(s,s'))$ resulting from executing the event model $((S',R',\mathsf{pre}),s')$ in the given epistemic state $(S,R,V),s$ is bisimilar to its refinement $((S',R',V'),s')$. The bisimulation $\mathfrak{R}'$ is as follows: all pairs $(t,\mathfrak{I}(u))$ in the restricted modal product are bisimilar to the state $u \in S'$ of which their second argument of the pair is the isomorphic image:

$$\mathfrak{R}'(t,\mathfrak{I}(u)) = u$$

Condition **atoms** is obvious, as refinement satisfies **atoms**. Condition **forth** is also obvious: if $((t, \mathfrak{I}(u)), (t', \mathfrak{I}(u'))) \in R_a''$, then by definition of the modal product $(\mathfrak{I}(u), \mathfrak{I}(u')) \in R_a'$, so $(u, u') \in R_a'$ (and, indeed, $((t', \mathfrak{I}(u')), u') \in \mathfrak{R}'$ by definition). Condition **back** is not obvious but also holds. Let $(u, u') \in R_a'$ and $((t, \mathfrak{I}(u)), u) \in \mathfrak{R}'$. There must be a $t' \in S$ (modulo bisimilarity) such that $(t', u') \in \mathfrak{R}$ so that $(t', \mathfrak{I}(u'))$ is in the modal product. We now have that from $(u, u') \in R_a'$ follows $(\mathfrak{I}(u), \mathfrak{I}(u')) \in R_a'$, and we also have that from $(u, u') \in R_a'$ follows $(t, t') \in R_a$ (as $\mathfrak{R}$ is a refinement). From $(t, t') \in R_a$ and $(\mathfrak{I}(u), \mathfrak{I}(u')) \in R_a'$ follows by definition the requested $(t, \mathfrak{I}(u)), (t', \mathfrak{I}(u')) \in R_a''$.

Lemma 1. *For every finite multi-agent epistemic model $M = (S, R, V)$, for every state $t \in S$ there exists a formula φ_t such that for all $u \in S$, $(M, u) \models \varphi_t$ if and only if (M, u) is bisimilar to (M, t).*

Proof. To show this it is sufficient to show that the relation

$$\mathfrak{R} = \{(v, w) \mid (M, v) \text{ and } (M, w) \text{ agree on the interpretation of all } \mathcal{L}_{el} \text{ formulae}\},$$

is a bisimulation. Clearly it satisfies **atoms**. To show that it satisfies **forth**, suppose $(v, w) \in \mathfrak{R}$ and let $a \leq m$, and let v^* be an a-successor of the state v. We note that the state w has finitely many a-successors, $w_1, \ldots, w_k$. Suppose for contradiction that none of these states were related to v^* by $\mathfrak{R}$. Thus for each such state w_a, there is a formula τ_a such that $(M, v^*) \not\models \tau_a$ and $(M, w) \models \tau_a$. However then we have that $(M, v) \models \neg K_a \left(\bigvee_{a=1}^k \tau_a \right)$, but $(M, w) \models K_a \left(\bigvee_{a=1}^k \tau_a \right)$, contradicting the fact that $(v, w) \in \mathfrak{R}$. Therefore, for every $a \leq m$, every a successor of v is related by $\mathfrak{R}$ to some a-successor of w. The property **back** can be shown symmetrically, so $\mathfrak{R}$ is a bisimulation.

This establishes that for every $u \in S$ where $(M, u) \not\leftrightarrow (M, t)$, there is some formula δ_t^u such that $(M, t) \models \delta_t^u$ and $(M, u) \not\models \delta_t^u$. We define $\varphi_t = \bigwedge \{\delta_t^u \mid u \in S \text{ and } (M, u) \not\leftrightarrow (M, t)\}$. Since bisimilar states agree on the interpretation of all $\mathcal{L}_{el}$ formulas, for all $u \in S$, $(M, u) \models \varphi_t$ if and only if $(M, u) \leftrightarrow (M, t)$.

We emphasize that the notion of event model relative to a language allows for *infinite* event models, unlike in a logic with an inductively defined language including (finite!) event models. That is to come next. This will also allow us to compare our proposal with a known method [5] for quantifying over events.

Definition 10 ([5]). *The language $\mathcal{L}_{aeml}$ of arbitrary event model logic is the language $\mathcal{L}_{fel}$ of future event logic with an additional inductive construct $[\mathsf{M}, \mathsf{s}]\varphi$.*

We can view $[\mathsf{M}, \mathsf{s}]\varphi$ as an inductive construct, because, given the (enumerable) set of event model frames, $[\mathsf{M}, \mathsf{s}]$ can be seen as an operation on $|\mathcal{D}(\mathsf{M})|$ arguments of type formula (similar to automata-PDL). These arguments are the preconditions of the events in the event model. The language $\mathcal{L}_{aeml}$ can also be seen as extension with construct $G\varphi$ of the language $\mathcal{L}_{eml}$ for event model logic.

To distinguish future event logic from logics with the same language but other semantics for $G\varphi$, we also write $\models_{\leftarrow}$ instead of $\models$ for the forcing relation in future event logic; we (always) write $\models_\otimes$ for the forcing relation in arbitrary event model logic.

For the semantics of $G\varphi$ in terms of event models we need to restrict the preconditions of their events to G-free formulas, i.e. $\mathcal{L}_{eml}$ formulas. This is to avoid circularity in the definition, as $G\varphi$ could itself be a precondition of such an event. An event model is G-free iff all preconditions of its events are G-free.

Definition 11 (Semantics of arbitrary event model logic). *Where the preconditions of events in any* M *are G-free.*

$$M, s \models_\otimes G\varphi \text{ iff for all } G\text{-free } (\mathsf{M}, \mathsf{s}) : M, s \models_\otimes [\mathsf{M}, \mathsf{s}]\varphi$$

There are refinements of epistemic models that cannot be seen as the result of executing an event model. This is because event models (in the language) have by definition a finite domain. For example, given a finite epistemic model (M, s), consider its unwinding as an infinite tree (representing the bisimulation class). This is a refinement of (M, s). But the result of executing a finite event model in a finite epistemic model cannot be an infinite tree. Of course, that tree is bisimilar to the initial epistemic state so can be seen in another sense as the result of execution the trivial event. But:

Because of the restriction to G-free preconditions in event models, we will still not get precise correspondence between the two semantics. The crux is that there are more epistemic distinctions in models than can be enumerated by epistemic formulas, see [5] for a similar matter. (However, we do not have a counterexample.)

Restricted to the class of finite epistemic models we still have that:

Proposition 6. *Let M be finite. Then:* $M, s \models_\leftrightarrow \varphi$ *iff* $M, s \models_\otimes \varphi$.

Proof. Directly from Propositions 5 and 4.

5   Bisimulation and Information

Instead of validating $F\varphi$ in some (M, s) by finding a refinement of (M, s), we can equivalently find a *model restriction of a bisimilar epistemic state*. This alternative semantics $\models_\leftrightarrow$ is interesting because of a relationship with bisimulation quantifiers [11], for which many theoretical results are known; and it is also interesting because it shows that every informative update is equivalent to public announcement of factual information "of which the agents may not have been aware."

Definition 12. *Below, S' is the domain of M', and S'' is such that $s \in S''$:*

$$M, s \models_\leftrightarrow G\varphi \text{ iff for all } (M', s') \leftrightarrow (M, s) \text{ and for all } S'' \subseteq S' : M'|S'', s' \models_\leftrightarrow \varphi$$

On first thought it might seem that there are more refinements of a given model than domain restrictions of bisimilar models. In a refinement we can both restrict the domain (remove states) and remove links between states (delete pairs of the accessibility relation for an agent). But removing links between states can also be seen as a domain restriction on a even larger bisimilar model.

Proposition 7. $M, s \models_{\leftarrow} \varphi$ iff $M, s \models_{\leftrightarrow} \varphi$.

Proof. This can be shown by induction on the complexity of formulas. As $\models_{\leftarrow}$ and $\models_{\leftrightarrow}$ agree on the interpretations of atoms and all operators except G, it is sufficient to show that given $(M, s \models_{\leftarrow} \varphi$ iff $M, s \models_{\leftrightarrow} \varphi)$ we have $(M, s \models_{\leftarrow} G\varphi$ iff $M, s \models_{\leftrightarrow} G\varphi)$. From left to right the latter is trivial, because the refinements of (M, s) include the bisimulations of (M, s). For the direction from right to left, it suffices to show that any refinement (M', s') of model (M, s) is the restriction of a model (M'', s'') that is bisimilar to (M, s). This model is constructed as follows:

Let $M = (S, R, V)$, $M' = (S', R', V')$, and suppose that the refinement relation is $\mathfrak{R}$. Consider $(M'', s'') = ((S \oplus S', R'', V''), (s', 1))$, where for all agents $a \in A$

$$((s, 0), (t, 0)) \in R''_a \;\; \text{iff} \;\; (s, t) \in R_a$$
$$((s', 1), (t', 1)) \in R''_a \;\; \text{iff} \;\; (s', t') \in R'_a$$
$$((s', 1), (t, 0)) \in R''_a \;\; \text{iff} \;\; \exists s \in S : (s, s') \in \mathfrak{R} \text{ and } (s, t) \in R_a$$

We can then define the relation $\mathfrak{R}'$ between (M, s) and $(M'', (s, 0))$ as follows:

$$(s, (s', 1)) \in \mathfrak{R}' \;\; \text{iff} \;\; (s, s') \in \mathfrak{R}$$
$$(s, (s, 0)) \in \mathfrak{R}' \;\; \text{iff} \;\; s \in S$$

This relation $\mathfrak{R}'$ is a bisimulation: it *still* satisfies **back** since the states of S added to M' also satisfy **back**: any relation between them copied their relation in the original M. But it now also satisfies **forth**:

If $(s, (s, 0)) \in \mathfrak{R}'$ and $(s, t) \in R_a$ then by definition of the first clause of $\mathfrak{R}'$ we have $(t, (t, 0)) \in \mathfrak{R}'$ and, trivially by the definition of R''_a we have $((s, 0), (t, 0)) \in R''_a$. If $(s, (s', 1)) \in \mathfrak{R}'$ and $(s, t) \in R_a$ then we have (as before) $(t, (t, 0)) \in \mathfrak{R}'$ and $((s', 1), (t, 0)) \in R''_a$. The latter holds because of the *third* clause in the definition of R''_a.

Since $M''|(S' \times \{1\})$ is isomorphic to M' this concludes the proof.

We proceed by explaining the stated relation of this semantics with bisimulation quantifiers.

6   Bisimulation Quantifiers

Suppose that apart from the atoms in P we had an additional, reserved, atom r. The future temporal operator can be seen as (existential) bisimulation quantification over r. This relation becomes clear if we consider the restricted bisimulation version of the semantics for F:

> First choose a bisimilar epistemic state, then do a model restriction in that epistemic state that contains the actual state.

Given the class of models also valuing r we can replace this by

> First choose a P-bisimilar epistemic state (but where the valuation of r may vary wildly), then do a model restriction in that epistemic state that contains the actual state.

Of course we can match the variation in the valuation of r, as long as it contains the actual state, with that model restriction so we get

> First choose a P-bisimilar epistemic state, then do a model restriction to the r-states in that epistemic state, on condition that it contains the actual state.

The part "choose a P-bisimilar epistemic state" of this informal description is the semantics of a existential bisimulation quantification.

Definition 13. *Where V' is the valuation of M'.*

$$M, s \models_{\forall r} G\varphi \text{ iff for all } (M', s') \underline{\leftrightarrow}_P (M, s) : s' \in V'(r) \text{ implies } M'|r, s' \models_{\forall r} \varphi$$

Example 2. For an example, consider again the model with common uncertainty about the value of an atom p for agents a and b, where p is true. We now operate on models that also value the atom r, in the figure this is the value of the second digit: note that r is *not* part of the logical language! Given the bisimulation quantification, the initial value of r does not matter. In this model the formula $F(K_a p \wedge \neg K_a K_b p)$ is true. The first transition is to a model that is bisimilar with respect to p only. The second transition is a restriction to the states where r is true.

$$
\begin{array}{ccccc}
 & & 01 - ab - 11 & & 01 - ab - 11 \\
 & & | \qquad\quad | & & | \\
 & & b \qquad\quad b & & b \\
 & & | \qquad\quad | & & | \\
00 - ab - \underline{10} \quad \Rightarrow & & 00 - ab - \underline{11} \quad \Rightarrow & & \underline{11}
\end{array}
$$

Proposition 8. $M, s \models_{\underline{\leftrightarrow}} \varphi$ iff $M, s \models_{\forall r} \varphi$

Corollary 1. *On finite models and given common knowledge in the language, the four different semantics for G correspond.* (I.e. $\models_{\underline{\leftrightarrow}}$, $\models_{\underline{\leftarrow}}$, $\models_{\otimes}$, and $\models_{\forall r}$.)

Note that the extra atom r does not disturb these results. As yet it is mere surplus luggage that we're carrying along towards the next section where it will become more meaningful. Our fourth perspective of bisimulation quantifier semantics is useful for theoretical and for practical reasons. A theoretical consequence is that

Proposition 9. *Future event logic is decidable.*

Proof. Consider some $\varphi \in \mathcal{L}_{fel}$. Replace all occurrences of G in φ by $\forall r[r]$. It is decidable whether $\varphi^{\forall r}$ is satisfiable. (The decidability of bisimulation quantified modal logics can be generalized to multi-agent logics. Note that it also holds for specific model classes such as $\mathcal{K}D45$, $\mathcal{S}5$ and the modal μ-calculus; see [14].)

This is a useful result. If we add dynamic event model operators to future event logic (the language $\mathcal{L}_{aeml}$) we obtain arbitrary event model logic (see Definition

10). The restriction of this arbitrary event model logic to events that are public announcements is the logic $APAL$ investigated in [5]. For that logic, the satisfiability problem is undecidable (see [15]). That result also motivated this current investigation, because it promised more decidable logics.

However, we may note that the translation given (replace all occurrences of G in φ by $\forall r[r]$) is an accurate translation for *all* logics that are closed under bisimulation quantifiers and announcement. From a recent result of van Benthem and Ikegami [14] we know that the modal μ-calculus is also closed under products with event models. Since future event logic and arbitrary event model logic agree on the interpretation of $G\varphi$ over finite models (Proposition 6), we can conclude that the satisfiability problem for $\mathcal{L}_{aeml}$ restricted to finite models is reducible to the satisfiability problem for the μ-calculus, and hence decidable.

Note that the G-operator in arbitrary event model logic is interpreted differently (see Definition 11), and it is unknown whether this logic is decidable.

Our current perspective also provides us with additional modelling insight, namely that every informative update corresponds to the public announcement of an atomic fact. Kind of. What kind of? So far, it is unclear how to interpret this new perspective: we compare semantics with respect to model classes for different sets of atomic propositions; we did not add the fresh atom r to the logical language $\mathcal{L}_{fel}$. Here is where some trouble seems to start. If we merely add r as a formula to the language, but, e.g., rule out $K_a r$, we cannot truly interpret a r-restriction of a model as a public announcement: what use is an announcement of r if we cannot express that an agent a knows r after its announcement? But if we add r as just another propositional variable to the base clause of our inductive language definition, we run into trouble of a different kind: an existential bisimulation quantification means that the value of p is scrambled. Even with the restriction that the value of r remains unchanged in the actual state, we may now still have that an agent a knew r before an event, but has forgotten it afterwards, or vice versa. This is highly undesirable!

Example 3. In the previous example, we have that initially agent b knows that r is false: $K_b \neg r$, but after the update he apparently has forgotten that: $\neg K_b \neg r$. For another example: $K_a r \rightarrow F \neg K_a r$ would be a validity.

A technical solution to this dilemma, that at least makes the public announcement clear, is to

> replace all occurrences of G in formulas by occurrences of $\forall r[r]$,

where $\forall r$ is universal bisimulation over r and where $[r]$ stands for public announcement of r. Public announcement is a singleton event model, accessible to all agents, where there precondition of the event is the formuma between brackets, in this case: r. If we also allow r as formula, we can now interpret formulas of form $[r]\varphi$ in the usual sense for such events. For example, in our running example it is initially true that $\exists r \langle r \rangle (K_a p \wedge \neg K_a K_b p)$, as this is the translation of $F(K_a p \wedge \neg K_a K_b p)$. (For $\neg[\varphi]\neg\psi$ we write $\langle\varphi\rangle\psi$.)

But the real solution to this seeming dilemma is to consider an existential bisimulation as "the agents become aware of an additional fact," about which uncertainty is possible. From a modelling point of view this means that, before the bisimulation operation, the value of r should be *don't care*, in other words, "the agents are unaware of r," and the bisimulation quantification itself then means "the agents become aware of r." This is now in the proper sense that we move to a bisimilar model except for atom r, and (unlike before!) without the restriction that r should remain true in the actual state, because maybe it was false in the first place. And after that it should be possible for them to know that r, or know that $\neg r$: they are now aware of their uncertainty about r. Of course after that, there might be other facts the agents might become aware of. If we merely add r to the base of the inductive language definition we cannot express this. We need one more step. That final step we will now set in the next section.

7   Becoming Aware of Factual Information

First, we add more structure: For each epistemic model M, the set of atoms P is the disjoint union of a set of *relevant facts* $P_r(M)$ and a set of *irrelevant facts* $P_i(M)$. The set of relevant facts is typically finite. Then, in a given model, the interpretation of formulas containing irrelevant facts is undefined, unless they are bound by a bisimulation quantifier: we can only interpret irrelevant facts *after* they have *become* relevant to the agents. The bisimulation quantifier makes a fact relevant: its interpretation involves removing it from the set of irrelevant facts and adding it to the set of relevant facts. The result of this is that the value of irrelevant facts in any model is now truly *don't care* from the perspective of the agents. But they can still reason about the consequences of new facts after they were to become relevant, i.e., after the agents were to become aware of those facts.

Definition 14. *The language $\mathcal{L}_{qel}$ of quantified event logic is inductively defined as*

$$\varphi ::= p \mid \neg\varphi \mid (\varphi \wedge \varphi) \mid K_a\varphi \mid [\mathsf{M}, \mathsf{s}]\varphi \mid \forall p\varphi$$

where $a \in A$ and $p \in P$.

For the dual $\exists p\varphi$, read, "(there exists a fact p such that) after the agents have become aware of p, φ." We emphasize that by "becoming aware of p" we do not mean "learning that p is true". In the information state resulting from becoming aware of p, the agents may know that p is true, or that p is false, or have any epistemic uncertainty about its value, e.g., they may not know whether p is true, or one agent may know but not another, etc.

Definition 15 (Semantics). *Assume an epistemic model $M = (S, R, V)$ for atoms $P = P_r(M) \cup P_i(M)$. The interpretation of $\varphi \in \mathcal{L}_{qel}$ is defined by induction. We only give the clauses for relevant atoms p and for $\forall p\varphi$. The interpretation of irrelevant atoms is undefined. In the clause for $\forall p\varphi$ it is required that (M', s') is such that $P_r(M') = P_r(M) + p$ and $P_i(M') = P_i(M) - p$.*

$M, s \models p$ *iff* $s \in V_p$ *where* $p \in P_r(M)$

$M, s \models \forall p\varphi$ *iff for all* $(M', s') \underline{\leftrightarrow}_{P-p}(M, s) : M', s' \models \varphi$ *where* $p \in P_i(M)$

We have not explored this version in greater detail yet. Unlike the logics with temporal operators and the proposal with a reserved atom r for bisimulation quantification, agents may in this logic become aware of several different facts. Just as in Fagin and Halpern's logic of awareness [16] agents' awareness is about atoms only (and not about complex formulas), unlike their approach our awareness is global for all states and does not vary along different states. The main difference is of course that we propose a way to model *becoming* aware, unlike the *being* aware in [16].

We think this logic may help modellers construct epistemic models in steps. In this logic, if we say that agent a is uncertain about p, and we represent this in the two-state epistemic model, this now means that the agent *only* is uncertain about p. The value of other atoms in that epistemic state is *don't care*: information on an additional fact q might become available later, we then simply construct a p-but-not-q bisimulation of this current epistemic state that represents the agents' current knowledge, that includes q. This is exactly the $\exists q$-operation! We close this section with a suitable illustration of this.

Example 4. Initially the agents are only uncertain about p. Then, they become aware of q: in fact, a knows the value of q but b doesn't. Finally, it is announced that $p \vee q$. In the resulting state, a knows that p but b does not know that. Initially, the formula $\exists r \langle p \vee q \rangle (K_a p \wedge \neg K_b K_a p)$ is true. Observe that the bisimulation quantification is in this example different from the subsequent announcement. We can only achieve our final information state by announcing a more complex formula than an atom, and not by the announcement of an atom.

$$
\begin{array}{ccccc}
& & 01 - ab - 11 & & 01 - ab - 11 \\
& & | \qquad\qquad | & & | \\
& & b \qquad\qquad b & & b \\
& & | \qquad\qquad | & & | \\
0 - ab - \underline{1} \quad \Rightarrow & & 00 - ab - \underline{10} \quad \Rightarrow & & \underline{10}
\end{array}
$$

Further research. We are currently further exploring the modelling of awareness using bisimulation quantification, as in this section. We are also investigating the axiomatization of the future event logic that forms the core of our contribution, including its model checking complexity (relative to different model classes, such as $S5$), and expressivity issues.

References

1. Baltag, A., Moss, L.: Logics for epistemic programs. Synthese 139, 165–224 (2004); Knowledge, Rationality & Action 1–60
2. van Ditmarsch, H., van der Hoek, W., Kooi, B.: Dynamic Epistemic Logic. Synthese Library, vol. 337. Springer, Heidelberg (2007)
3. van Benthem, J., van Eijck, J., Kooi, B.: Logics of communication and change. Information and Computation 204(11), 1620–1662 (2006)

4. Hoshi, T.: The logic of communication graphs for group communication and the dynamic epistemic logic with a future operator. Philosophy Department, Stanford University (2006)
5. Balbiani, P., Baltag, A., van Ditmarsch, H., Herzig, A., Hoshi, T., Lima, T.D.: What can we achieve by arbitrary announcements? A dynamic take on Fitch's knowability. In: Samet, D. (ed.) Proceedings of TARK XI, Louvain-la-Neuve, Belgium, pp. 42–51. Presses Universitaires de Louvain (2007)
6. Balbiani, P., Baltag, A., van Ditmarsch, H., Herzig, A., Hoshi, T., Lima, T.D.: 'knowable' as 'known after an announcement'. Review of Symbolic Logic 1(3), 305–334 (2008)
7. van Benthem, J., Gerbrandy, J., Pacuit, E.: Merging frameworks for interaction: DEL and ETL. In: Samet, D. (ed.) Proceedings of TARK 2007, pp. 72–81 (2007)
8. Fagin, R., Halpern, J., Moses, Y., Vardi, M.: Reasoning about Knowledge. MIT Press, Cambridge (1995)
9. Lomuscio, A., Ryan, M.: An algorithmic approach to knowledge evolution. Artificial Intelligence for Engineering Design, Analysis and Manufacturing (AIEDAM) 13(2) (1998); Special issue on Temporal Logic in Engineering
10. Aczel, P.: Non-Well-Founded Sets. CSLI Lecture Notes, vol. 14. CSLI Publications, Stanford (1988)
11. Visser, A.: Bisimulations, model descriptions and propositional quantifiers. Logic Group Preprint Series 161, Department of Philosophy, Utrecht University (1996)
12. Hollenberg, M.: Logic and bisimulation. PhD thesis, University of Utrecht (1998)
13. French, T.: Bisimulation quantifiers for modal logic. PhD thesis, University of Western Australia (2006)
14. van Benthem, J., Ikegami, D.: Modal fixed-point logic and changing models. In: Avron, A., Dershowitz, N., Rabinovich, A. (eds.) Pillars of Computer Science. LNCS, vol. 4800, pp. 146–165. Springer, Heidelberg (2008); Also available as ILLC Prepublication Series PP-2008-19
15. French, T., van Ditmarsch, H.: Undecidability for arbitrary public announcement logic. In: Proceedings of the seventh conference "Advances in Modal Logic", London, pp. 23–42. College Publications (2008)
16. Fagin, R., Halpern, J.: Belief, awareness, and limited reasoning. Artificial Intelligence 34(1), 39–76 (1988)

On the Dynamics of Institutional Agreements

Andreas Herzig[1], Tiago de Lima[2], and Emiliano Lorini[1]

[1] IRIT, Toulouse, France
Andreas.Herzig@irit.fr,
Emiliano.Lorini@irit.fr
[2] Eindhoven University of Technology, The Netherlands
T.d.Lima@tue.nl

Abstract. In this work we continue the work initiated in [1], in which a logic of individual and collective acceptance was introduced. Our aim in this paper is to investigate the extension of the logic of acceptance by *public announcements* of formulas. The function of public announcements is to diminish the space of possible worlds accepted by agents and sets of agents while functioning as members of a given group, team, organization, institution, etc., x. If a set of agents C ends up with an empty set of worlds that they accept while functioning as members of x, then the agents in C do not identify themselves any longer with x. In such a situation the agents in C should have the possibility to join x again. To that aim we discuss at the end of the paper an operation which consists of an agent (or set of agents) joining a given group, team, organization, institution, etc.

1 Introduction

The concept of *collective acceptance* has been studied in social philosophy in opposition to group attitudes such as *common belief* and *common knowledge* that are popular in artificial intelligence and theoretical computer science [2,3]. As suggested in [4], the main difference between collective acceptance and common belief (or common knowledge) is that a collective acceptance by a set of agents C is based on the fact that the agents in C identify and recognize themselves as members of the same *social context*, such as a group, team, organization, institution, etc. Common belief (and common knowledge) does not necessarily entail this aspect of mutual recognition and identification with respect to a social context. In this sense, according to [4,5], collective acceptance rather than common belief is the more appropriate concept to characterize a proper notion of *group belief*. For example, in the context of the organization Greenpeace the agents in a set C (collectively) accept that their mission is to protect the Earth *qua* members of Greenpeace. The state of acceptance *qua* members of Greenpeace is the kind of acceptance the agents in C are committed to when they are functioning together as members of Greenpeace.

It has been emphasized that a similar distinction between acceptance and belief exists at the *individual* level. While an agent's belief that p is an attitude

J.-J.Ch. Meyer and J.M. Broersen (Eds.): KRAMAS 2008, LNAI 5605, pp. 66–80, 2009.

of the agent constitutively aimed at the truth of p, an agent's acceptance is not necessarily connected to the actual truth of the proposition. In order to better distinguish these two notions, it has been suggested in [6] that while an agent's beliefs are not subject to the agent's will, its acceptances are voluntary; while its beliefs aim at truth, its acceptances are sensitive to pragmatic considerations; while its beliefs are shaped by evidence, its acceptances need not be; finally, while its beliefs are context-independent, its acceptances might depend on context. Often the acceptances of an agent depend on social contexts, that is, while identifying itself as a member of a group (or team, organization, institution, etc.) an agent reasons and accepts things *qua* member of this group. In these situations it may happen that the agent's acceptances are in conflict with its beliefs. For instance, a lawyer who is trying to defend a client in a murder case accepts *qua* lawyer that the client is innocent, even if he believes the contrary.

The aim of this paper is to continue the work initiated in [1,7]. There, a logic of individual and collective acceptance was introduced.[1] One of the notable features of that logic is that the accessibility relation associated to the acceptance operator is not necessarily serial: an empty set of possible worlds associated to a group C in a context x just means that C does not identify itself with x.

Our aim here is to investigate the extension of the logic of acceptance by *public announcements* of formulas, noted $x!\psi$. Modal operators of type $[x!\varphi]$ are intended to express that the members of a certain group, team, organization, institution, etc., x learn that φ is true in that institution in such a way that their acceptances, *qua* members of x, are updated. The function of public announcements is to diminish the space of possible worlds accepted by agents and groups of agents. It might also happen that a given set of agents C ends up with an empty set of possible worlds that they accept while functioning as members of a certain social context x. As we have said, this means that C quits x: the agents in C do not identify themselves any longer with x. In such a situation C should have the possibility to join x again. To that aim we discuss at the end of the paper an operation which consists of an agent (or set of agents) joining a given social context x.

The main contribution of this paper is to extend the logic presented in [1] to public announcements and show that, differently from common belief and common knowledge, reduction axioms can be given. As usual, the addition of these axioms to the Hilbert axiomatics of the logic of acceptance provides a complete axiomatization of the logic of acceptance and announcements. As far as we know, this constitutes the first attempt to build up a logic of acceptance and public announcements.

The paper is organized as follows. In Section 2 we present the syntax and semantics of acceptance logic together with its axiomatization. In Section 3 we extend it with announcements, and show that our extension also allows for reduction axioms and thereby a complete axiomatization. In Section 4 we formalize an example which illustrates the dynamics of acceptance based on announcements.

---

[1] This logic has some similarities with the logic of *group belief* that we have developed in [8,9].

In Section 5 we briefly discuss the operation which consists of an agent (or set of agents) joining a social context. This section is not intended to provide a solution to the logical characterization of this social phenomenon though. In Section 6 we draw conclusions.

2 The Logic of Acceptance $\mathcal{AL}$

We now present a variant of the *Acceptance Logic* ($\mathcal{AL}$) that was introduced in [1]. $\mathcal{AL}$ enables expressing that certain agents identify themselves as members of a social context x and, reasoning about what agents and groups of agents accept while functioning together as members of a certain social context. The axioms of $\mathcal{AL}$ clarify the relationships between individual acceptance (acceptances of individual agents) and collective acceptance (acceptances of groups of agents).

2.1 Syntax

The syntactic primitives of $\mathcal{AL}$ are the following: a finite non-empty set of agents AGT; a countable set of atomic formulas ATM; and a finite set of labels $CTXT$ denoting social contexts such as groups, teams, organizations, institutions, etc. The language $\mathcal{L}_{\mathcal{AL}}$ of the logic $\mathcal{AL}$ is given by the following BNF:

$$\varphi ::= p \mid \neg\varphi \mid \varphi \vee \varphi \mid \mathcal{A}_{C:x}\varphi$$

where p ranges over ATM, C ranges over 2^{AGT}, and x ranges over $CTXT$. The formula $\mathcal{A}_{C:x}\varphi$ reads "the agents in C accept that φ while functioning together as members of x". We write $i{:}x$ instead of $\{i\}{:}x$.

For example, $\mathcal{A}_{C:Greenpeace}protectEarth$ expresses that the agents in C accept that the mission of Greenpeace is to protect the Earth while functioning as activists in the context of Greenpeace; and $\mathcal{A}_{i:Catholic}PopeInfallibility$ expresses that agent i accepts that the Pope is infallible while functioning as a Catholic in the context of the Catholic Church.

The intuition is that in two different contexts the same agent may accept contradictory propositions. For example, while functioning as a Catholic, agent i accepts that killing is forbidden, and while functioning as a soldier i accepts that killing is allowed. The CEO of Airbus accepts that Airbus is in good health while functioning as a member of Airbus Industries, and privately accepts the contrary.

The classical boolean connectives $\wedge$, $\rightarrow$, $\leftrightarrow$, $\top$ (tautology) and $\bot$ (contradiction) are defined from $\vee$ and $\neg$ in the usual manner.

The formula $\mathcal{A}_{C:x}\bot$ has to be read "agents in C are not functioning together as members of x", because we assume that functioning as a group member is, at least in this minimal sense, a rational activity. Conversely, $\neg\mathcal{A}_{C:x}\bot$ has to be read "agents in C are functioning together as members of x". Thus, $\neg\mathcal{A}_{C:x}\bot \wedge \mathcal{A}_{C:x}\varphi$ stands for "agents in C are functioning together as members of x and they accept that φ while functioning together as members of x" or simply "agents in C accept that φ *qua* members of x". This is a case of *group acceptance*. For the individual case, formula $\neg\mathcal{A}_{i:x}\bot \wedge \mathcal{A}_{i:x}\varphi$ has to be read "agent i accepts that φ *qua* member of x". This is a case of *individual acceptance*.

2.2   Semantics and Axiomatization

We use a standard possible worlds semantics. Let the set of all couples of non-empty subsets of agents and social contexts be

$$\Delta = \big\{ C{:}x \ : \ C \in 2^{AGT} \text{ and } x \in CTXT \big\}.$$

An *acceptance model* is a triple $\mathcal{M} = \langle W, \mathscr{A}, \mathscr{V} \rangle$ where:

- W is a non-empty set of possible worlds;
- $\mathscr{A} : \Delta \to W \times W$ maps every $C{:}x \in \Delta$ to a relation $\mathscr{A}_{C:x}$ between possible worlds in W; and
- $\mathscr{V} : ATM \to 2^{W}$ is valuation function associating a set of possible worlds $\mathscr{V}(p) \subseteq W$ to each atomic formula p of ATM.

We write $\mathscr{A}_{C:x}(w)$ for the set $\{ w' \ : \ \langle w, w' \rangle \in \mathscr{A}_{C:x} \}$. $\mathscr{A}_{C:x}(w)$ is the set of worlds that is accepted by the agents in C while functioning together as members of x.

Given $\mathcal{M} = \langle W, \mathscr{A}, \mathscr{V} \rangle$ and $w \in W$, the couple $\langle \mathcal{M}, w \rangle$ is a *pointed acceptance model*. The satisfaction relation $\models$ between formulas of $\mathcal{L_{AL}}$ and pointed acceptance models $\langle \mathcal{M}, w \rangle$ is defined as usual for atomic propositions, negation and disjunction. The satisfaction relation for acceptance operators is the following:

$$\mathcal{M}, w \models \mathcal{A}_{C:x}\varphi \quad \text{iff} \quad \mathcal{M}, w' \models \varphi \text{ for every } w' \in \mathscr{A}_{C:x}(w)$$

Validity of a formula φ (noted $\models_{\mathcal{AL}} \varphi$) is defined as usual.

The axiomatization of $\mathcal{AL}$ is given in Fig. 1. It is not meant to reflect the semantics the way it has been presented up to this point. Instead, these axioms are meant as postulates, and we are going to present the corresponding semantic constraints later.

As usual, the K-principles are the axioms and inference rules of the basic modal logic K. Axioms **4*** and **5*** are introspection axioms: when the agents in a set C function together as members of x, then, for all $y \in CTXT$ and B such that $B \subseteq C$, the agents in B have access to all the facts that are accepted (or that are not accepted) by the agents in C. In particular, if the agents in C (do not) accept that φ while functioning together as members of x then, while functioning together as members of y, the agents of every subset B of C accept that agents in C (do not) accept that φ.

(K)	All K-principles for the operators $\mathcal{A}_{C:x}$
(4*)	$\mathcal{A}_{C:x}\varphi \to \mathcal{A}_{B:y}\mathcal{A}_{C:x}\varphi$, if $B \subseteq C$
(5*)	$\neg\mathcal{A}_{C:x}\varphi \to \mathcal{A}_{B:y}\neg\mathcal{A}_{C:x}\varphi$, if $B \subseteq C$
(Inc)	$(\neg\mathcal{A}_{C:x}\bot \wedge \mathcal{A}_{C:x}\varphi) \to \mathcal{A}_{B:x}\varphi$, if $B \subseteq C$
(Una)	$\mathcal{A}_{C:x}(\bigwedge_{i \in C} \mathcal{A}_{i:x}\varphi \to \varphi)$

Fig. 1. Axiomatization of $\mathcal{AL}$

Example 1. Suppose that three agents i, j, k, while functioning together as members of the UK trade union, accept that their mission is to try to increase teachers' wages, but they do not accept *qua* members of the trade union that their mission is to try to increase railway workers' wages: $\mathcal{A}_{\{i,j,k\}:Union} IncrTeacherWage$ and $\neg\mathcal{A}_{\{i,j,k\}:Union} IncrRailwayWage$. By axiom **4*** we infer that, while functioning as a UK citizen, i accepts that i, j, k accept that their mission is to try to increase teachers' wages, while functioning together as members of the trade union: $\mathcal{A}_{i:UK}\mathcal{A}_{\{i,j,k\}:Union} IncrTeacherWage$. By Axiom **5*** we infer that, while functioning as a UK citizen, i accepts that i, j, k do not accept, *qua* members of the trade union, that their mission is to try to increase railway workers' wages: $\mathcal{A}_{i:UK}\neg\mathcal{A}_{\{i,j,k\}:Union} IncrRailwayWage$.

Axiom **Inc** says that, if the agents in C accept that φ *qua* members of x then every subset B of C accepts φ while functioning together as members of x. This means that things accepted by the agents in C *qua* members of a certain social context x are necessarily accepted by agents in all of C's subsets with respect to the same context x. Axiom **Inc** describes the *top down* process leading from C's collective acceptance to the individual acceptances of C's members.[2]

Example 2. Imagine three agents i, j, k that, *qua* players of the game Clue, accept that someone called Mrs. Red, has been killed:
$$\neg\mathcal{A}_{\{i,j,k\}:Clue}\bot \wedge \mathcal{A}_{\{i,j,k\}:Clue} killedMrsRed.$$
By axiom **Inc** we infer that also the two agents i, j, while functioning as Clue players, accept that someone called Mrs. Red has been killed:
$$\mathcal{A}_{\{i,j\}:Clue} killedMrsRed.$$

Axiom **Una** expresses a unanimity principle according to which the agents in C, while functioning together as members of x, accept that if each of them individually accepts that φ while functioning as member of x, then φ is the case. This axiom describes the *bottom up* process leading from individual acceptances of the members of C to the collective acceptance of the group C.

In order to make our axioms valid we impose the following constraints on acceptance models, for any world $w \in W$, context $x, y \in CTXT$, and coalitions $C, B \in 2^{AGT}$ such that $B \subseteq C$:

(S.1) if $w' \in \mathscr{A}_{B:y}(w)$ then $\mathscr{A}_{C:x}(w') = \mathscr{A}_{C:x}(w)$;
(S.2) if $\mathscr{A}_{C:x}(w) \neq \emptyset$ then $\mathscr{A}_{B:x}(w) \subseteq \mathscr{A}_{C:x}(w)$;
(S.3) if $w' \in \mathscr{A}_{C:x}(w)$ then $w' \in \bigcup_{i \in C} \mathscr{A}_{i:x}(w')$.

Axioms **4*** and **5*** together correspond to the constraint **S.1**; axiom **Inc** corresponds to **S.2**, and axiom **Una** to **S.3** (in the sense of correspondence theory). As all our axioms are in the Sahlqvist class we obtain straightforwardly: .

---

[2] Note that the more general
$$(\neg\mathcal{A}_{C:x}\bot \wedge \mathcal{A}_{C:x}\varphi) \to \mathcal{A}_{B:y}\varphi, \text{ if } B \subseteq C$$
would lead to unwanted consequences: the group of Catholics' acceptance *qua* members of the Catholic church that the Pope is infallible does not entail that Catholics privately accept that the Pope is infallible.

Also note that for $B \subseteq C$, neither $\mathcal{A}_{C:x}\bot \to \mathcal{A}_{B:x}\bot$ nor $\mathcal{A}_{B:x}\bot \to \mathcal{A}_{C:x}\bot$ should hold.

Theorem 1. *The axiomatization of $\mathcal{AL}$ of Fig. 1 is sound and complete w.r.t. the class of $\mathcal{AL}$ models satisfying constraints **S.1**, **S.2**, and **S.3**.*

Proof. It is a routine task to check that all the axioms of the logic $\mathcal{AL}$ correspond to their semantic counterparts. It is routine, too, to check that all $\mathcal{AL}$ axioms are in the Sahlqvist class, for which a general completeness result exists [10]. □

Example 3. It follows from axioms **4***, **5*** and **Inc** that if $B \subseteq C$ then $\models_{\mathcal{AL}}$ $\mathcal{A}_{B:y}\mathcal{A}_{C:x}\varphi \leftrightarrow \mathcal{A}_{B:y}\bot \vee \mathcal{A}_{C:x}\varphi$ and $\models_{\mathcal{AL}} \mathcal{A}_{B:y}\neg\mathcal{A}_{C:x}\varphi \leftrightarrow \mathcal{A}_{B:y}\bot \vee \neg\mathcal{A}_{C:x}\varphi$. We also have $\models_{\mathcal{AL}} \mathcal{A}_{C:x}(\mathcal{A}_{C:x}\varphi \rightarrow \varphi)$.

3   The Logic of Acceptance and Public Announcements $\mathcal{ALA}$

In its nature, acceptance comes by communication: if a group accepts that one of its members i accepts that φ then this is often the result of a speech act performed by i. Acceptance is therefore closely related to the notion of commitment that has been studied in agent communication languages [11,12,13].

   In this paper we study the combination of acceptance logic $\mathcal{AL}$ with a rather simple communicative act, viz. public announcements as defined in public announcement logic ($\mathcal{PAL}$) [14]. Basically, when formula ψ is publicly announced, all agents learn that ψ is true. Our truth condition is that of Kooi [15], that is slightly different from the standard one in public announcement logic: it does not require announcements to be truthful.

3.1   Language and Models

The language $\mathcal{L}_{\mathcal{ALA}}$ of acceptance logic with announcements ($\mathcal{ALA}$) extends $\mathcal{L}_{\mathcal{AL}}$ by modal formulas of the form $[x!\psi]\varphi$, where $\varphi, \psi \in \mathcal{L}_{\mathcal{AL}}$. Such formulas are read "φ holds after the public announcement of ψ in context x". Modal operators of type $[x!\psi]$ are intended to express that the agents learn that ψ is true in the social context x, in such a way that their acceptances, *qua* members of x, are updated.

   The announcement $x!\mathcal{A}_{i:x}\psi$ is an *event*. It approximates i's *action* of announcing that ψ in context x. (This is an assertion in speech act theory and in Walton and Krabbe's dialogue games [16].)

   It is worth noting that when x denotes an institution, events of type $x!\psi$ can be used to describe the event of issuing or promulgating a certain norm ψ (e.g. obligation, permission) within the context of the institution x.[3]

   Formulas of $\mathcal{L}_{\mathcal{ALA}}$ are interpreted in pointed acceptance models. The satisfaction relation $\models$ of Section 2 is extended by the following clause:

$$\langle W, \mathscr{A}, \mathscr{V} \rangle, w \models [x!\psi]\varphi \quad \text{iff} \quad \langle W, \mathscr{A}^{x!\psi}, \mathscr{V} \rangle, w \models \varphi$$

---

[3] For a logical characterization of the act of *proclaiming* or *promulgating* a norm, see also [17].

with

- $\mathscr{A}_{C:y}^{x!\psi}(w) = \mathscr{A}_{C:y}(w)$, for all $C{:}y \in \Delta$, $w \in W$ and $y \neq x$;
- $\mathscr{A}_{C:y}^{x!\psi}(w) = \mathscr{A}_{C:y}(w) \cap ||\psi||_{\mathcal{M}}$, for all $C{:}y \in \Delta$, $w \in W$ and $y = x$,

where as usual $||\psi||_{\mathcal{M}} = \{w \ : \ \mathcal{M}, w \models \psi\}$ is the extension of ψ in $\mathcal{M}$, i.e., the set of worlds where ψ is true. Thus, in a way similar to [15], the agents take into account the announcement of ψ in the social context x and modify their acceptances *qua* members of x by eliminating all arrows leading to $\neg\psi$ worlds (instead of eliminating the worlds themselves, as in $\mathcal{PAL}$). On the contrary, when x and y are different, the accessibility relations associated to the acceptances *qua* members of y are not modified, after the announcement of ψ in the social context x.

Validity of a formula φ (noted $\models_{\mathcal{ALA}} \varphi$) is defined as before. For example, $\models_{\mathcal{ALA}} [x!p]\mathcal{A}_{C:x}p$, and $\models_{\mathcal{ALA}} \mathcal{A}_{C:x}\neg p \to [x!p]\mathcal{A}_{C:x}\bot$. The latter means that coalition C quits all social contexts within which C's acceptances are inconsistent with what is announced.

Note that contrarily to standard common knowledge and belief, the modified accessibility relations for acceptances are not computed from the modified accessibility relations for individuals, but are first-class citizens here: they are changed 'on their own'.

Proposition 1. *If $\mathcal{M}$ is an acceptance model then $\mathcal{M}^{x!\psi}$ is an acceptance model.*

Proof. We show that $\mathcal{M}^{x!\psi} = \langle W, \mathscr{A}^{x!\psi}, \mathscr{V} \rangle$ satisfies **S.1**, **S.2** and **S.3**. In what follows let $B \subseteq C$.

(S.1): Let $w_2 \in \mathscr{A}_{B:y}^{x!\psi}(w_1)$. If the latter is true then $w_2 \in \mathscr{A}_{B:y}(w_1)$, which implies $\mathscr{A}_{C:x}(w_2) = \mathscr{A}_{C:x}(w_1)$, because $\mathcal{M}$ respects **S.1**.
Now, we show that $\mathscr{A}_{C:x}^{x!\psi}(w_2) \subseteq \mathscr{A}_{C:x}^{x!\psi}(w_1)$. Consider a possible world $w_3 \in \mathscr{A}_{C:x}^{x!\psi}(w_2)$. This means that $w_3 \in \mathscr{A}_{C:x}(w_2) \cap ||\psi||_{\mathcal{M}}$. Then, in particular, $w_3 \in \mathscr{A}_{C:x}(w_2)$, which implies $w_3 \in \mathscr{A}_{C:x}^{x!\psi}(w_1)$, because $\mathscr{A}_{C:x}(w_2) = \mathscr{A}_{C:x}(w_1)$. By using an analogous argument, we show that $\mathscr{A}_{C:x}^{x!\psi}(w_1) \subseteq \mathscr{A}_{C:x}^{x!\psi}(w_2)$.

(S.2): Let $\mathscr{A}_{C:x}^{x!\psi}(w_1) \neq \emptyset$ and $w_2 \in \mathscr{A}_{B:x}^{x!\psi}(w_1)$. We show that $w_2 \in \mathscr{A}_{C:x}^{x!\psi}(w_1)$. The hypothesis implies $w_2 \in \mathscr{A}_{B:x}(w_1) \cap ||\psi||_{\mathcal{M}}$. Then, in particular, $w_2 \in \mathscr{A}_{B:x}(w_1)$. Also note that the hypothesis implies $\mathscr{A}_{C:x}(w_1) \neq \emptyset$. Then, $w_2 \in \mathscr{A}_{C:x}(w_1)$, because $\mathcal{M}$ respects **S.2**. We conclude that $w_2 \in \mathscr{A}_{C:x}(w_1) \cap ||\psi||_{\mathcal{M}}$. The latter is true if and only if $w_2 \in \mathscr{A}_{C:x}^{x!\psi}(w_1)$.

(S.3): Let $w_2 \in \mathscr{A}_{C:x}^{x!\psi}(w_1)$. We show that $w_2 \in \mathscr{A}_{i:x}^{x!\psi}(w_2)$ for some $i \in C$. The hypothesis is equivalent to $w_2 \in \mathscr{A}_{C:x}(w_1) \cap ||\psi||_{\mathcal{M}}$. Then, in particular, $w_2 \in \mathscr{A}_{C:x}(w_1)$, which implies $w_2 \in \mathscr{A}_{i:x}(w_2)$ for some $i \in C$, because $\mathcal{M}$ respects **S.3**. Then, $w_2 \in \mathscr{A}_{i:x}(w_2) \cap ||\psi||_{\mathcal{M}}$ for some $i \in C$. The latter is true if and only if $w_2 \in \mathscr{A}_{i:x}^{x!\psi}(w_2)$ for some $i \in C$. $\qquad\square$

3.2   Reduction Axioms and Completeness

Just as in dynamic epistemic logics without common belief, $\mathcal{ALA}$ has reduction axioms for all cases (individual and collective acceptance). This contrasts with logics having the common belief operator, for which such axioms do not exist [18].

Proposition 2. *The following equivalences are* $\mathcal{ALA}$ *valid.*

(R.1) $[x!\psi]p \leftrightarrow p$
(R.2) $[x!\psi]\neg\varphi \leftrightarrow \neg[x!\psi]\varphi$
(R.3) $[x!\psi](\varphi_1 \wedge \varphi_2) \leftrightarrow [x!\psi]\varphi_1 \wedge [\psi!\varphi]_2$
(R.4) $[x!\psi]\mathcal{A}_{C:y}\varphi \leftrightarrow \mathcal{A}_{C:y}[x!\psi]\varphi$ *(if $y \neq x$)*
(R.5) $[x!\psi]\mathcal{A}_{C:y}\varphi \leftrightarrow \mathcal{A}_{C:y}(\psi \rightarrow [x!\psi]\varphi)$ *(if $y = x$)*

Proof. **(R.1)**:
$\quad \langle W, \mathscr{A}, \mathscr{V}\rangle, w \models [x!\psi]p$
$\quad$ iff $\langle W, \mathscr{A}^{x!\psi}, \mathscr{V}\rangle \models p$
$\quad$ iff $w \in \mathscr{V}(p)$
$\quad$ iff $\langle W, \mathscr{A}, \mathscr{V}\rangle, w \models p$.

(R.2):
$\quad \langle W, \mathscr{A}, \mathscr{V}\rangle, w \models [x!\psi]\neg\varphi$
$\quad$ iff $\langle W, \mathscr{A}^{x!\psi}, \mathscr{V}\rangle, w \models \neg\varphi$
$\quad$ iff $\langle W, \mathscr{A}^{x!\psi}, \mathscr{V}\rangle, w \not\models \varphi$
$\quad$ iff $\langle W, \mathscr{A}, \mathscr{V}\rangle, w \not\models [x!\psi]\varphi$
$\quad$ iff $\langle W, \mathscr{A}, \mathscr{V}\rangle, w \models \neg[x!\psi]\varphi$.

(R.3):
$\quad \langle W, \mathscr{A}, \mathscr{V}\rangle, w \models [x!\psi](\varphi_1 \wedge \varphi_2)$
$\quad$ iff $\langle W, \mathscr{A}^{x!\psi}, \mathscr{V}\rangle, w \models \varphi_1 \wedge \varphi_2$
$\quad$ iff $\langle W, \mathscr{A}^{x!\psi}, \mathscr{V}\rangle, w \models \varphi_1$ and $\langle W, \mathscr{A}^{x!\psi}, \mathscr{V}\rangle, w \models \varphi_2$
$\quad$ iff $\langle W, \mathscr{A}, \mathscr{V}\rangle, w \models [x!\psi]\varphi_1$ and $\langle W, \mathscr{A}, \mathscr{V}\rangle, w \models [x!\psi]\varphi_2$
$\quad$ iff $\langle W, \mathscr{A}, \mathscr{V}\rangle, w \models [x!\psi]\varphi_1 \wedge [x!\psi]\varphi_2$.

(R.4): We show that the equivalent $\neg[x!\psi]\mathcal{A}_{C:y}\varphi \leftrightarrow \neg\mathcal{A}_{C:y}[x!\psi]\varphi$ is valid.
$\quad \langle W, \mathscr{A}, \mathscr{V}\rangle, w \models \neg[x!\psi]\mathcal{A}_{C:y}\varphi$
$\quad$ iff $\langle W, \mathscr{A}, \mathscr{V}\rangle, w \models [x!\psi]\neg\mathcal{A}_{C:y}\varphi$, by **(R.3)**,
$\quad$ iff $\langle W, \mathscr{A}^{x!\psi}, \mathscr{V}\rangle, w \models \neg\mathcal{A}_{C:y}\varphi$
$\quad$ iff there is $w' \in \mathscr{A}_{C:y}^{x!\psi}(w)$ such that $\langle W, \mathscr{A}^{x!\psi}, \mathscr{V}\rangle, w' \models \neg\varphi$
$\quad$ iff there is $w' \in \mathscr{A}_{C:y}(w)$ such that $\langle W, \mathscr{A}^{x!\psi}, \mathscr{V}\rangle \models \neg\varphi$, because $y \neq x$,
$\quad$ iff there is $w' \in \mathscr{A}_{C:y}(w)$ such that $\langle W, \mathscr{A}, \mathscr{V}\rangle \models [x!\psi]\neg\varphi$
$\quad$ iff there is $w' \in \mathscr{A}_{C:y}(w)$ such that $\langle W, \mathscr{A}, \mathscr{V}\rangle \models \neg[x!\psi]\varphi$, by **(R.3)**,
$\quad$ iff $\langle W, \mathscr{A}, \mathscr{V}\rangle \models \neg\mathcal{A}_{C:y}[x!\psi]\varphi$.

(R.5): We show that $\neg[x!\psi]\mathcal{A}_{C:y}\varphi \leftrightarrow \neg\mathcal{A}_{C:y}(\psi \rightarrow [x!\psi]\varphi)$ is valid.
$\quad \langle W, \mathscr{A}, \mathscr{V}\rangle, w \models \neg[x!\psi]\mathcal{A}_{C:y}\varphi$
$\quad$ iff $\langle W, \mathscr{A}, \mathscr{V}\rangle, w \models [x!\psi]\neg\mathcal{A}_{C:y}\varphi$, by **(R.3)**,
$\quad$ iff $\langle W, \mathscr{A}^{x!\psi}, \mathscr{V}\rangle, w \models \neg\mathcal{A}_{C:y}\varphi$
$\quad$ iff there is $w' \in \mathscr{A}_{C:y}^{x!\psi}(w)$ such that $\langle W, \mathscr{A}^{x!\psi}, \mathscr{V}\rangle, w' \models \neg\varphi$
$\quad$ iff there is $w' \in \mathscr{A}_{C:y}(w)$ such that $\langle W, \mathscr{A}, \mathscr{V}\rangle \models \psi$ and $\langle W, \mathscr{A}^{x!\psi}, \mathscr{V}\rangle \models \neg\varphi$, because $y = x$,
$\quad$ iff there is $w' \in \mathscr{A}_{C:y}(w)$ such that $\langle W, \mathscr{A}, \mathscr{V}\rangle \models \psi$ and $\langle W, \mathscr{A}, \mathscr{V}\rangle \models [x!\psi]\neg\varphi$
$\quad$ iff there is $w' \in \mathscr{A}_{C:y}(w)$ such that $\langle W, \mathscr{A}, \mathscr{V}\rangle \models \psi \wedge [x!\psi]\neg\varphi$
$\quad$ iff there is $w' \in \mathscr{A}_{C:y}(w)$ such that $\langle W, \mathscr{A}, \mathscr{V}\rangle \models \psi \wedge \neg[x!\psi]\varphi$, by **(R.3)**,
$\quad$ iff $\langle W, \mathscr{A}, \mathscr{V}\rangle \models \neg\mathcal{A}_{C:y}(\psi \rightarrow [x!\psi]\varphi)$. $\square$

These equivalences are called reduction axioms because they allow to rewrite every formula by successively eliminating the announcement operators, ending up with a formula that contains none.

Theorem 2. *For every $\mathcal{ALA}$ formula there is an equivalent $\mathcal{AL}$ formula.*

Proof. The proof goes just as for public announcement logic (found in [15]): each of the above $\mathcal{ALA}$ valid equivalences **R.2**–**R.5**, when applied from the left to the right, yields a simpler formula, where 'simpler' roughly speaking means that the announcement operator is pushed inwards. Once the announcement operator attains an atom it is eliminated by the first equivalence **R.1**. $\square$

Theorem 3. *The formulas that are valid in $\mathcal{ALA}$ models are completely axiomatized by the axioms and inference rules of $\mathcal{AL}$ together with the reduction axioms of Proposition 2.*

Proof. This a straightforward consequence of Theorem 1 and Theorem 2. $\square$

Here are some examples of reductions.

Example 4. The formula $[x!p]\mathcal{A}_{C:x}p$ is successively rewritten as follows:
$$\mathcal{A}_{C:x}(p \to [x!p]) \qquad\qquad\qquad\qquad\qquad\qquad \text{by } \mathbf{R.5}$$
$$\mathcal{A}_{C:x}(p \to p) \qquad\qquad\qquad\qquad\qquad\qquad\qquad \text{by } \mathbf{R.1}$$
The latter is a theorem of every normal modal logic (and therefore also of acceptance logic $\mathcal{AL}$). It follows that the initial formula is valid, too.

Example 5. The formula $\mathcal{A}_{i:x}\neg p \to [x!p]\mathcal{A}_{i:x}\bot$ is rewritten as follows:
$$\mathcal{A}_{i:x}\neg p \to \mathcal{A}_{i:x}(p \to [x!p]\bot) \qquad\qquad\qquad\qquad \text{by } \mathbf{R.5}$$
$$\mathcal{A}_{i:x}\neg p \to \mathcal{A}_{i:x}(p \to \bot) \qquad\qquad\qquad\qquad\qquad \text{by } \mathbf{R.1}$$
The latter is a theorem of every normal modal logic (and therefore also of acceptance logic $\mathcal{AL}$).

Example 6. The formula $[x!(\mathcal{A}_{i:x}p)]\mathcal{A}_{C:x}\mathcal{A}_{i:x}p$ is rewritten as follows:
$$\mathcal{A}_{C:x}(\mathcal{A}_{i:x}p \to [x!(\mathcal{A}_{i:x}p)]\mathcal{A}_{i:x}p) \qquad\qquad\qquad \text{by } \mathbf{R.5}$$
$$\mathcal{A}_{C:x}(\mathcal{A}_{i:x}p \to \mathcal{A}_{i:x}(\mathcal{A}_{i:x}p \to [x!(\mathcal{A}_{i:x}p)]p)) \qquad\quad \text{by } \mathbf{R.5}$$
$$\mathcal{A}_{C:x}(\mathcal{A}_{i:x}p \to \mathcal{A}_{i:x}(\mathcal{A}_{i:x}p \to p)) \qquad\qquad\qquad\quad \text{by } \mathbf{R.1}$$
The latter is an $\mathcal{AL}$ theorem (because $\mathcal{A}_{i:x}(\mathcal{A}_{i:x}p \to p)$ is an $\mathcal{AL}$ theorem, cf. Example 3 of Section 2). It follows that the initial formula is valid, too.

3.3   Discussion

As said before, $\mathcal{ALA}$ has reduction axioms for all cases (individual and collective acceptance), while it has been shown in [18] that logics of common belief (and common knowledge) do not. Technically, it happens because acceptance models have more arrows. Let $\mathcal{M}$ be the epistemic model in Fig. 2. Note that even though $\mathcal{M}, w \not\models \mathbf{C}_{ij}(q \to [!q]p)$, we still have $\mathcal{M}, w \models [!q]\mathbf{C}_{ij}p$. In words, it is not common belief among i and j that q implies that after the public announcement of q, we have that p, but after the public announcement of q it is common belief among i and j that p. That is, common belief may appear 'out of the blue', i.e., it was not foreseeable by the agents, it just 'pops up'.

However, when we build the acceptance model that corresponds to the model $\mathcal{M}$, it looks like the structure in Fig. 3. Semantic constraint **S.1** obliges the corresponding acceptance model to have more arrows, in particular, it must have the dashed arrow from w to u labelled by $ij{:}x$. Let $\mathcal{M}'$ be such acceptance model, note that $\mathcal{M}', w \not\models \mathbf{A}_{ij:x}(q \rightarrow [!q]p)$, and also $\mathcal{M}', w \not\models [x!q]\mathbf{A}_{ij:x}p$. That is, contrary to common belief, common acceptances cannot just 'pop up' without be foreseeable by the agents.

None the less, we argue that the reduction axiom **R.5** is an intuitive property of collective acceptance. This is due to the fact that, differently from the standard notions of common belief and common knowledge, collective acceptance entails an aspect of mutual identification and recognition with respect to a group.

Consider the left to right direction of the reduction axiom **R.5**. When the agents in a set C identify themselves with a group x and recognize each other as members of this group, they accept certain rules and principles to stand for the the rules and principles of the group. That is, the agents in C *share a common body* of rules and principles. Among these shared rules and principles, there are the rules and principles which describe how the world should evolve when an announcement occurs. They govern how the acceptance of the agents in the group will be changed after an announcement. Suppose that a certain fact ψ is publicly announced. After this announcement, the agents in C accept φ, while identifying themselves with a group x and recognizing each other as members of this group: $[x!\psi]\mathcal{A}_{C:x}\varphi$. This collective acceptance of the agents in C is not created from scratch after the announcement of ψ. On the contrary, the creation of this acceptance depends on what the agents in C accepted (before the announcement) as a principle of group x. In particular, the creation of C's acceptance that φ rests on the fact that, before ψ is announced, the agents in C, while identifying themselves and recognizing each other as members of x, accept a principle saying that "if ψ is true then, after ψ is announced in x, φ will be true": $\mathcal{A}_{C:x}(\psi \rightarrow [x!\psi]\varphi)$.

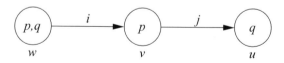

Fig. 2. An epistemic model

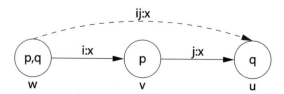

Fig. 3. The acceptance model corresponding to the epistemic model in Fig. 2

For example, imagine that the agents in a set C identify themselves and recognize each other as members of the Lilliputian pacifist movement. Let ψ denote the proposition "the government of Lilliput has decided to attack the neighboring nation of Blefuscu".[4] After ψ is publicly announced the agents in C accept that $\varphi =$ "they should start to protest against the Lilliput government", while functioning as members of the Lilliputian pacifist movement: $[LilliputPacifist!\psi]\mathcal{A}_{C:LilliputPacifist}\varphi$. This implies that (before the announcement) the agents in C, while identifying themselves and recognizing each other as members of the Lilliputian pacifist movement, accept a principle saying that "if ψ is true then, after ψ is announced, φ will be true": $\mathcal{A}_{C:LilliputPacifist}(\psi \rightarrow [LilliputPacifist!\psi]\varphi)$. That is, the creation of C's acceptance to protest against the Lilliput government depends on the fact that, before the announcement, the agents in C accept to protest against the Lilliput government in case it will announce its decision to attack the neighboring nation of Blefuscu. This means that C's acceptance to protest depends on the fact that, before the announcement, the agents in C accept a principle which specifies what to do in case the Lilliput government will manifest its intention to attack Blefuscu.

4   An Example

Until now we only considered that group acceptances emerge from consensus, by admitting axiom **Una**. One can go further and also consider other kinds of group acceptances, as shown in the next example. The example is inspired by Pettit [19].

Example 7. Imagine a three-member court which has to make a judgment on whether a defendant is liable (noted l) for a breach of contract. The three judges i, j and k accept a majority rule to decide on the issue. That is, i, j and k, while functioning as members of the court, accept that if the majority of them accepts that the defendant is liable (resp. not liable), then the defendant is liable (resp. not liable). Formally, for any B such that $B \subseteq \{i, j, k\}$ and $|B| = 2$ we have:

$$(\textbf{Maj})\quad \mathcal{A}_{\{i,j,k\}:court}(\bigwedge_{i\in B} \mathcal{A}_{i:court}l \rightarrow l) \quad \wedge \quad \mathcal{A}_{\{i,j,k\}:court}(\bigwedge_{i\in B} \mathcal{A}_{i:court}\neg l \rightarrow \neg l)$$

Given the previous majority rule, we can prove that: after the announcement that both i and j accept l (the defendant is liable) while functioning as members of the court, the agents in $\{i, j, k\}$ accept l while functioning together as members of the court. Indeed, from the previous majority rule we can derive the formula $[court!\mathcal{A}_{i:court}l \wedge \mathcal{A}_{j:court}l]\mathcal{A}_{\{i,j,k\}:court}l$. To prove this, it is sufficient to note that, by means of the reduction axioms, the formula $[court!\mathcal{A}_{i:court}l \wedge \mathcal{A}_{j:court}l]\mathcal{A}_{\{i,j,k\}:court}l$ is successively rewritten as follows:

---

[4] Lilliput and Blefuscu are the two fictional nations, permanently at war, that appear in the novel "Gulliver's Travels" by Jonathan Swift.

$$\mathcal{A}_{\{i,j,k\}:court}((\mathcal{A}_{i:court}l \wedge \mathcal{A}_{j:court}l) \rightarrow [court!\mathcal{A}_{i:court}l \wedge \mathcal{A}_{j:court}l]l) \qquad \text{by } \mathbf{R.5}$$
$$\mathcal{A}_{\{i,j,k\}:court}((\mathcal{A}_{i:court}l \wedge \mathcal{A}_{j:court}l) \rightarrow l) \qquad \text{by } \mathbf{R.1}$$

The latter is entailed by the majority rule **Maj**.

In the previous example, we have considered a majority rule as a principle which is responsible for the creation of collective acceptances from individual acceptances. This is stronger than the basic axiom of unanimity (**Una**) of $\mathcal{AL}$. One can imagine other kinds of rules. For instance, one can consider social contexts with leaders (see also [7]). In such contexts, one can formalize the rule according to which everything that the leaders accept is universally accepted in the social context. Let the set of leaders of x be $L_x \in 2^{AGT}$. Then one can formalize that everything that the leaders accept is universally accepted in the social context by:

$$\textbf{(Leader)} \quad \mathcal{A}_{C:x}(\mathcal{A}_{L_x:x}\varphi \rightarrow \varphi)$$

5 Adding Retractions to $\mathcal{ALA}$: Some General Insights

According to our semantics, $\mathcal{A}_{i:x}\neg p \rightarrow [x!p]\mathcal{A}_{i:x}\bot$ is an $\mathcal{ALA}$ theorem (cf. Example 5). In words, when p is publicly announced then i quits all contexts x where he accepted p: agent i is no longer part of the institution, is kicked out of the group, etc. In $\mathcal{ALA}$ there is no means for i to get out of that situation and re-integrate context x. At the present stage, our logic of acceptance does not include an operation which consists of an agent (or set of agents) joining a certain social context.

Semantically, what we need is the opposite of the previous model restrictions: an operation of adding arrows labelled by $i{:}x$ to the model. Syntactically, what we need is a new form of announcements $i \leftarrow C{:}x$ and corresponding modal operators of type $[i \leftarrow C{:}x]$, meaning that agent i adopts C's acceptances in context x. In terms of Kripke models, the accessibility relation $\mathcal{A}_{i:x}$ is identified with $\mathcal{A}_{C:x}$. This kind operation of adding arrows is reminiscent of the logic of preference upgrade of van Benthem and Liu [20], and the logic of granting and revoking permissions of Pucella and Weissman [21].[5] More intuitively, $i \leftarrow C{:}x$ represents the operation of agent i's joining the social context x by adopting the acceptances of group C of members of x. After this operation, agent i should start to function again as members of x.

Other kinds of retraction operations can be devised. For example, one might want to consider the operation of creating a supergroup D of a given group C, where D takes over all of C's acceptances. The logical form of such an operation might be expressed by the operator $[D{:}{=}C:x]$. This operation should allow in particular to express that the agents in D start to function as members of x (i.e., to move from $\mathcal{A}_{D:x}\bot$ to $\neg\mathcal{A}_{D:x}\bot$), by taking over all acceptances of the agents in the subgroup C.

---

[5] See [22] for a systemic study of these operators.

We are currently working on the technical issue of providing a semantic characterization and axiomatics of the previous operations $i \leftarrow C{:}x$ and $D{:=}C{:}x$ and corresponding modal operators $[i \leftarrow C{:}x]$ and $[D{:=}C{:}x]$.

6 Conclusion

In this paper we continued the studies initiated in [1], where the logic $\mathcal{AL}$, intended to formalize group (and individual) acceptances, was proposed. Here we extend $\mathcal{AL}$ by public announcements. As far as we know, our approach es novel and there is no other attempt to build up a logic of acceptance and public announcements.

The public announcement of ψ is an event that results in all agents learning that ψ is true. The public announcement of $\mathcal{A}_{C{:}x}\psi$ can be understood as a speech act. It simulates the announcement made by the group C itself, that they accept ψ while functioning as members of x. Therefore, as seen in Example 7, public announcements can be used to reason about the acceptances of agents when they express their own acceptances to each other. For instance, in that particular example we saw that a public announcement makes one of the agents quit the group, since he learns that the acceptances of the other agents are contrary to his own acceptances in the same context. As noted in Section 3.1, when the social context x denotes an institution, announcements of the form $x!\psi$ can be used to describe the event of issuing or promulgating a certain norm ψ (e.g. obligation, permission) within the context of the institution x.

We also provide a complete axiomatization for the logic of acceptances and announcements $\mathcal{ALA}$. As well as for epistemic logic with public announcements, the axiomatization given for $\mathcal{ALA}$ uses reduction axioms. In $\mathcal{ALA}$, group acceptances are related to individual acceptances, but they are not computed from them. It contrasts with epistemic logics where the concept of common knowledge (or common belief) is completely defined in terms of individual knowledge (or belief). Due to this difference, it is possible to have reduction axioms for group acceptances, while it is known to be impossible for common knowledge. Still, in Section 3.3 we argue that this is an intuitive feature of group acceptances.

Acknowledgements

We would like to thank the three anonymous reviewers for their very helpful comments.

The contribution by Andreas Herzig and Emiliano Lorini was supported by the French ANR project 'ForTrust: Social Trust Analysis and Formalization'. The contribution by Tiago de Lima is part of the research program Moral Responsibility in R&D Networks, supported by the Netherlands Organisation for Scientific Research (NWO), under grant number 360-20-160.

References

1. Gaudou, B., Longin, D., Lorini, E., Tummolini, L.: Anchoring institutions in agents' attitudes: Towards a logical framework for autonomous multi-agent systems. In: Padgham, L., Parkes, D.C. (eds.) Proceedings of AAMAS 2008, pp. 728–735. IFAAMAS (2008)
2. Fagin, R., Halpern, J., Moses, Y., Vardi, M.: Reasoning about Knowledge. The MIT Press, Cambridge (1995)
3. Lewis, D.K.: Convention: a philosophical study. Harvard University Press, Cambridge (1969)
4. Gilbert, M.: On Social Facts. Routledge, London (1989)
5. Tuomela, R.: The Philosophy of Sociality. Oxford University Press, Oxford (2007)
6. Hakli, P.: Group beliefs and the distinction between belief and acceptance. Cognitive Systems Research (7), 286–297 (2006)
7. Lorini, E., Longin, D.: A logical approach to institutional dynamics: from acceptances to norms via legislators. In: Brewka, G., Lang, J. (eds.) Proceedings of KR 2008. AAAI Press, Menlo Park (2008)
8. Gaudou, B., Herzig, A., Longin, D.: Grounding and the expression of belief. In: Proceedings of KR 2006, pp. 211–229. AAAI Press, Menlo Park (2006)
9. Gaudou, B., Herzig, A., Longin, D., Nickles, M.: A new semantics for the FIPA agent communication language based on social attitudes. In: Brewka, G., Coradeschi, S., Perini, A., Traverso, P. (eds.) Proceedings of ECAI 2006, pp. 245–249. IOS Press, Amsterdam (2006)
10. Blackburn, P., de Rijke, M., Venema, Y.: Modal Logic. Cambridge University Press, Cambridge (2001)
11. Fornara, N., Colombetti, M.: Operational specification of a commitment-based agent communication language. In: Castelfranchi, C., Johnson, W.L. (eds.) Proceedings of AAMAS 2002, Bologna, pp. 535–542. ACM Press, New York (2002)
12. Verdicchio, M., Colombetti, M.: A Logical Model of Social Commitment for Agent Communication. In: Proceedings of AAMAS 2003, pp. 528–535. ACM, New York (2003)
13. Singh, M.P.: Agent communication languages: Rethinking the principles. IEEE Computer 31(12), 40–47 (1998)
14. Plaza, J.: Logics of public communications. In: Emrich, M.L., Hadzikadic, M., Pfeifer, M.S., Ras, Z.W. (eds.) Proceedings of ISMIS 1989, pp. 201–216 (1989)
15. Kooi, B.: Expressivity and completeness for public update logic via reduction axioms. Journal of Applied Non-Classical Logics 17(2), 231–253 (2007)
16. Walton, D.N., Krabbe, E.C.: Commitment in Dialogue: Basic Concepts of Interpersonal Reasoning. State University of New-York Press (1995)
17. Gelati, J., Rotolo, A., Sartor, G., Governatori, G.: Normative autonomy and normative co-ordination: Declarative power, representation, and mandate. Artificial Intelligence and Law 12(1-2), 53–81 (2004)
18. Kooi, B., Van Benthem, J.: Reduction axioms for epistemic actions. In: Schmidt, R., Pratt-Hartmann, I., Reynolds, M., Wansing, H. (eds.) Proceedings AiML 2004, pp. 197–211. King's College Publications (2004)
19. Pettit, P.: Deliberative democracy and the discursive dilemma. Philosophical Issues 11, 268–299 (2001)

20. van Benthem, J., Liu, F.: Dynamic logic of preference upgrade. Journal of Applied Non-Classical Logics 17(2), 157–182 (2007)
21. Pucella, R., Weissman, V.: Reasoning about dynamic policies. In: Walukiewicz, I. (ed.) FOSSACS 2004. LNCS, vol. 2987, pp. 453–467. Springer, Heidelberg (2004)
22. Aucher, G., Balbiani, P., Farias Del Cerro, L., Herzig, A.: Global and local graph modifiers. In: Methods for Modalities 5 (M4M-5). ENTCS. Elsevier, Amsterdam (2007)

From Trust in Information Sources to Trust in Communication Systems: An Analysis in Modal Logic

Emiliano Lorini and Robert Demolombe

Institut de Recherche en Informatique de Toulouse (IRIT), France
Emiliano.Lorini@irit.fr
Robert.Demolombe@irit.fr

Abstract. We present a logical analysis of trust that integrates in the definition of trust: the truster's goal and the truster's belief that the trustee has the right properties (powers, abilities, dispositions) to ensure that the goal will be achieved. The second part of the paper is focused on the specific domain of trust in information sources and communication systems. We provide an analysis of the properties of information sources (validity, completeness, sincerity and cooperativity) and communication systems (availability and privacy) and, we discuss their relationships with trust.

1 Introduction

Future computer applications such as the semantic Web [3], e-business and e-commerce [16], Web services [28] will be open distributed systems in which the many constituent components are agents spread throughout a network in a decentralized manner. These agents will interact between them in flexible ways in order to achieve their design objectives and to accomplish the tasks which are delegated to them by the human users. Some of them will directly interact and communicate with the human users. During the system's lifetime, these agents will need to manage and deal with trust. They will need to automatically make trust judgments in order to assess the trustworthiness of other (software and human) agents while, for example, exchanging money for a service, giving access to a certain information, choosing between conflicting sources of information. They will also need to understand how trust can be induced in a human user in order to support his interaction with the system and to motivate him to use the application. Consequently, these agents will need to understand the components and the determinants of the user's trust in the system.

Thus, to realize all their potential, future computer applications will require the development of sophisticated formal and computational models of trust. These models must provide clear definitions of the relevant concepts related to trust and safe reasoning rules which can be exploited by the agents for assessing the trustworthiness of a given target. Moreover, these models of trust must be cognitively plausible, so that they can be directly exploited by the agents during their interactions with the human user in order to induce him to trust the system and the underlying Information and Communication Technology (ICT) infrastructure. With cognitively plausible models of trust, we mean models in which the main cognitive constituents of trust as a mental attitude are identified (*e.g.* beliefs, goals).

J.-J.Ch. Meyer and J.M. Broersen (Eds.): KRAMAS 2008, LNAI 5605, pp. 81–98, 2009.

This paper follows our previous works [25,20] with the objective of developing a general formal model of trust which meets the previous desiderata. It is worth noting that it is not our aim to propose a model of trust based on statistics about past interactions with a given target and reputational information. In particular, the present paper focuses on an issue that we have neglected up to now: the issue of trust in information sources and communication systems. We think that this issue is very relevant for future computer applications such as the semantic Web, e-business and Web services. For example, in a typical scenario of e-business, trust in information sources has a strong influence on an agent's decision to buy, or to sell, a specific kind of stocks. Indeed, to take such a decision an agent has several types of information sources to consult in order to predict the future evolution of the stock value. These information sources may be banks, companies, consultants, *etc.* and the agent may believe that some of these information sources have a good competence but are not necessarily sincere, others are reluctant to inform about bad news, others are competent but are not necessarily informed at the right moment, *etc.* In a typical scenario of Web services, an agent might want to make a credit card transaction by means of a certain online payment system. In this case, the agent's trust in the communication system has a strong influence on the agent's decision to exploit it for the credit card transaction. In particular, the agent's trust in the online payment system is supported by the agent's belief that the online payment system will ensure the privacy of the credit card number from potential intruders.

The paper is organized as follows. We start with a presentation of a modal logic which enables reasoning about actions, beliefs and goals of agents (Section 2). This logic will be used during the paper for formalizing the relevant concepts of our model of trust. Then, a general definition of trust is presented (Section 3). Section 4 is focused on the formal characterization of some important properties of an information source: validity, completeness, sincerity and cooperativity. In Section 5 we show that these properties are epistemic supports for trust in information sources. In Section 6 we provide an analysis of communication systems. We define two fundamental properties of communication systems: availability and privacy. Then, in Section 7, we show that these properties are epistemic supports for an agent's trust in a communication system. We conclude with a discussion of some related works and we show some directions for future works.

2 A Modal Logic of Beliefs, Goals and Actions

We present in this section the multimodal logic $\mathcal{L}$ that we use in the paper to formalize the relevant concepts of our model of trust. $\mathcal{L}$ combines the expressiveness of a dynamic logic [19] with the expressiveness of a logic of agents' mental attitudes [9,29].

2.1 Syntax and Semantics

The syntactic primitives of the logic $\mathcal{L}$ are the following: a nonempty finite set of agents $AGT = \{i, j, \ldots\}$; a nonempty finite set of atomic actions $AT = \{a, b, \ldots\}$; a finite set of atomic formulas $\Pi = \{p, q, \ldots\}$. LIT is the set of literals which includes all atomic formulas and their negations, that is, $LIT = \{p, \neg p | p \in \Pi\}$. We note $P, Q, \ldots$ the elements in LIT. We also introduce specific actions of the form

$inf_j(P)$ denoting the action of informing agent j that P is true. We call them informative actions. The set $INFO$ of informative actions is defined as follows: $INFO = \{inf_j(P)|j \in AGT, P \in LIT\}$. Since the set Π is finite, the set $INFO$ is finite as well. The set ACT of complex actions is given by the union of the set of atomic actions and the set of informative actions, that is: $ACT = AT \cup INFO$. We note $\alpha, \beta, \ldots$ the elements in ACT. The language of $\mathcal{L}$ is the set of formulas defined by the following BNF:

$$\varphi ::= p \mid \neg\varphi \mid \varphi \vee \varphi \mid \mathtt{After}_{i:\alpha}\varphi \mid \mathtt{Does}_{i:\alpha}\varphi \mid \mathtt{Bel}_i\varphi \mid \mathtt{Goal}_i\varphi$$

where p ranges over Π, α ranges over ACT and i ranges over AGT.

The operators of our logic have the following intuitive meaning. $\mathtt{Bel}_i\varphi$ is meant to stand for "agent i believes that φ"; $\mathtt{After}_{i:\alpha}\varphi$ is meant to stand for "after agent i does α, it is the case that φ" ($\mathtt{After}_{i:\alpha}\bot$ means "agent i cannot do action α"); $\mathtt{Does}_{i:\alpha}\varphi$ is meant to stand for "agent i is going to do α and φ will be true afterward" ($\mathtt{Does}_{i:\alpha}\top$ means "agent i is going to do α"); $\mathtt{Goal}_i\varphi$ is meant to stand for "agent i has the chosen goal that φ" (which can be shortened to "agent i wants φ to be true"). Note that operators $\mathtt{Goal}_i$ are used to denote an agent's chosen goals, that is, the goals that the agent has decided to pursue. We do not consider how an agent's chosen goals originate through deliberation from more primitive motivational attitudes called desires (see e.g.[30] on this issue). Since an agent's chosen goals result from the agent's deliberation, they must satisfy two fundamental rationality principles: chosen goals have to be consistent (i.e., a rational agent cannot decide to pursue inconsistent state of affairs); chosen goals have to be compatible with the agent's beliefs (i.e., a rational agent cannot decide to pursue something that it believes to be impossible). These two principles will be formally expressed in Section 4.

The following abbreviations are given:

$$\mathtt{Can}_i(\alpha) \stackrel{def}{=} \neg\mathtt{After}_{i:\alpha}\bot;$$
$$\mathtt{Int}_i(\alpha) \stackrel{def}{=} \mathtt{Goal}_i\mathtt{Does}_{i:\alpha}\top;$$
$$\mathtt{Inf}_{i,j}(P) \stackrel{def}{=} \mathtt{Does}_{i:inf_j(P)}\top.$$

$\mathtt{Can}_i(\alpha)$ means "agent i can do action α" (i.e. "i has the capacity to do α"). $\mathtt{Int}_i(\alpha)$ means "agent i intends to do α". Finally, $\mathtt{Inf}_{i,j}(P)$ means "i informs j that P is true". Models of the logic $\mathcal{L}$ are tuples $M = \langle W, R, D, B, G, V \rangle$ defined as follows.

- W is a non empty set of possible worlds or states.
- $R : AGT \times ACT \longrightarrow W \times W$ maps every agent i and action α to a relation $R_{i:\alpha}$ between possible worlds in W.
- $D : AGT \times ACT \longrightarrow W \times W$ maps every agent i and action α to a deterministic relation $D_{i:\alpha}$ between possible worlds in W.[1]
- $B : AGT \longrightarrow W \times W$ maps every agent i to a serial, transitive and Euclidean[2] relation B_i between possible worlds in W.

---

[1] A relation $D_{i:\alpha}$ is deterministic iff, if $(w, w') \in D_{i:\alpha}$ and $(w, w'') \in D_{i:\alpha}$ then $w' = w''$.
[2] A relation B_i on W is Euclidean if and only if, if $(w, w') \in B_i$ and $(w, w'') \in B_i$ then $(w', w'') \in B_i$.

- $G : AGT \longrightarrow W \times W$ maps every agent i to a serial relation G_i between possible worlds in W.
- $V : W \longrightarrow 2^{\Pi}$ is a truth assignment which associates each world w with the set $V(w)$ of atomic propositions true in w.

It is convenient to view relations on W as functions from W to 2^W; therefore we write $D_{i:\alpha}(w)$ for the set $\{w' : (w, w') \in D_{i:\alpha}\}$, *etc.* Given a world $w \in W$, $B_i(w)$ is the set of worlds which are compatible with agent i's beliefs at w and $G_i(w)$ is the set of worlds which are compatible with agent i's goals at w.

Given a world $w \in W$, $R_{i:\alpha}(w)$ is the set of worlds that can be reached from w through the occurrence of agent i's action α; $D_{i:\alpha}(w)$ is the *next* world of w which will be reached from w through the occurrence of agent i's action α. Indeed, we have two kinds of relations for specifying the dynamic dimension of frames:

- when $D_{i:\alpha}(w) = \{w'\}$ then at w agent i performs an action α resulting in the next state w';
- when $w' \in R_{i:\alpha}(w)$ but $w' \notin D_{i:\alpha}(w)$ then if at w agent i would do something different from what it actually does it might have produced another outcome world w'.

If $D_{i:\alpha}(w) \neq \emptyset$ (resp. $R_{i:\alpha}(w) \neq \emptyset$) then, we say that $D_{i:\alpha}$ (resp. $R_{i:\alpha}$) is defined at w. Note that the relation $D_{i:\alpha}$ is used here to specify that an action α is going to performed by a certain agent i and, consequently, that an agent i wants that a certain action α is going to be performed by him (i.e. agent i intends to perform action α).

Given a model M, a world w and a formula φ, we write $M, w \models \varphi$ to mean that φ is true at world w in M. The rules defining the truth conditions of formulas are just standard for atomic formulas, negation and disjunction. The following are the remaining truth conditions for $\text{After}_{i:\alpha}\varphi$, $\text{Does}_{i:\alpha}\varphi$, $\text{Bel}_i\varphi$ and $\text{Goal}_i\varphi$:

- $M, w \models \text{After}_{i:\alpha}\varphi$ iff $M, w' \models \varphi$ for all w' such that $(w, w') \in R_{i:\alpha}$,
- $M, w \models \text{Does}_{i:\alpha}\varphi$ iff $\exists w'$ such that $(w, w') \in D_{i:\alpha}$ and $M, w' \models \varphi$,
- $M, w \models \text{Bel}_i\varphi$ iff $M, w' \models \varphi$ for all w' such that $(w, w') \in B_i$,
- $M, w \models \text{Goal}_i\varphi$ iff $M, w' \models \varphi$ for all w' such that $(w, w') \in G_i$.

Observe that the modal operator $\text{Does}_{i:\alpha}$ is of type possibility, and that all other modal operators are of type necessity.

The following section is devoted to illustrate the additional semantic constraints over $\mathcal{L}$ models and a sound and complete axiomatization of the logic $\mathcal{L}$.

2.2 Semantic Constraints and Axiomatization

The axiomatization of the logic $\mathcal{L}$ includes all tautologies of propositional calculus and the rule of inference *modus ponens* (**MP**).

MP　　From $\vdash \varphi$ and $\vdash \varphi \rightarrow \psi$ infer $\vdash \psi$

Operators for actions of type $\text{After}_{i:\alpha}$ and $\text{Does}_{i:\alpha}$ are normal modal operators satisfying the axioms and rules of inference of system K. Operators of type Bel_i and Goal_i are just standard normal modal operators. The former are standard doxastic operators satisfying the axioms and rules of inference of system KD45, whereas the latter are modal operators for goal similar to the operators introduced in [9] satisfying the axioms and rules of inference of system KD. That is, the following axioms and rules of inference for every operator Bel_i, Goal_i, $\text{After}_{i:\alpha}$ and $\text{Does}_{i:\alpha}$ are given.

$\mathbf{K}_{Bel}$	$(\text{Bel}_i\varphi \wedge \text{Bel}_i(\varphi \to \psi)) \to \text{Bel}_i\psi$
$\mathbf{K}_{Goal}$	$(\text{Goal}_i\varphi \wedge \text{Goal}_i(\varphi \to \psi)) \to \text{Goal}_i\psi$
$\mathbf{K}_{After}$	$(\text{After}_{i:\alpha}\varphi \wedge \text{After}_{i:\alpha}(\varphi \to \psi)) \to \text{After}_{i:\alpha}\psi$
$\mathbf{K}_{Does}$	$(\text{Does}_{i:\alpha}\varphi \wedge \neg\text{Does}_{i:\alpha}\neg\psi) \to \text{Does}_{i:\alpha}(\varphi \wedge \psi)$
$\mathbf{D}_{Bel}$	$\neg(\text{Bel}_i\varphi \wedge \text{Bel}_i\neg\varphi)$
$\mathbf{D}_{Goal}$	$\neg(\text{Goal}_i\varphi \wedge \text{Goal}_i\neg\varphi)$
$\mathbf{4}_{Bel}$	$\text{Bel}_i\varphi \to \text{Bel}_i\text{Bel}_i\varphi$
$\mathbf{5}_{Bel}$	$\neg\text{Bel}_i\varphi \to \text{Bel}_i\neg\text{Bel}_i\varphi$
$\mathbf{Nec}_{Bel}$	From $\vdash \varphi$ infer $\vdash \text{Bel}_i\varphi$
$\mathbf{Nec}_{Goal}$	From $\vdash \varphi$ infer $\vdash \text{Goal}_i\varphi$
$\mathbf{Nec}_{After}$	From $\vdash \varphi$ infer $\vdash \text{After}_{i:\alpha}\varphi$
$\mathbf{Nec}_{Does}$	From $\vdash \varphi$ infer $\vdash \neg\text{Does}_{i:\alpha}\neg\varphi$

Actions and intentions. We add the following constraint over every relation $D_{i:\alpha}$ and every relation $D_{j:\beta}$ of all $\mathcal{L}$ models. For every $i, j \in AGT$, $\alpha, \beta \in ACT$ and $w \in W$:

S1   if $D_{i:\alpha}$ and $D_{j:\beta}$ are defined at w then $D_{i:\alpha}(w) = D_{j:\beta}(w)$.

Constraint $S1$ says that if w' is the *next* world of w which is reachable from w through the occurrence of agent i's action α and w'' is also the *next* world of w which is reachable from w through the occurrence of agent j's action β, then w' and w'' denote the same world. Indeed, we suppose that every world can only have one *next* world. Note that $S1$ implies determinism of every $D_{i:\alpha}$. The semantic constraint $S1$ corresponds (in the sense of correspondence theory, see *e.g.*[5]) to the following Axiom $\mathbf{Alt}_{Does}$.

$\mathbf{Alt}_{Does}$ $\text{Does}_{i:\alpha}\varphi \to \neg\text{Does}_{j:\beta}\neg\varphi$

Axiom $\mathbf{Alt}_{Does}$ says that: if i is going to do α and φ will be true afterward, then it cannot be the case that j is going to do β and $\neg\varphi$ will be true afterward. We also suppose that the world is never static in our framework, that is, we suppose that for every world w there exists some agent i and action α such that i is going to perform α at w. Formally, for every $w \in W$ we have that:

S2   there exist $i \in AGT$ and $\alpha \in ACT$ such that $D_{i:\alpha}$ is defined at w.

The semantic constraint $S2$ corresponds to the following axiom of our logic.

$\mathbf{Active}$ $\bigvee_{i \in AGT, \alpha \in ACT} \text{Does}_{i:\alpha}\top$

Axiom $\mathbf{Active}$ ensures that for every world w there is a *next* world of w which is reachable from w by the occurrence of some action of some agent. This is the reason why the operator X for *next* of LTL (linear temporal logic) can be defined as follows:[3]

---

[3] Note that X satisfies the standard property $\text{X}\varphi \leftrightarrow \neg\text{X}\neg\varphi$ (i.e. φ will be true in the next state iff $\neg\varphi$ will not be true in the next state).

$$X\varphi \stackrel{\text{def}}{=} \bigvee_{i \in AGT, \alpha \in ACT} \text{Does}_{i:\alpha}\varphi.$$

The following relationship is supposed between every relation $D_{i:\alpha}$ and the corresponding relation $R_{i:\alpha}$ of all $\mathcal{L}$ models. For every $i \in AGT$, $\alpha \in ACT$ and $w \in W$:

S3 $D_{i:\alpha} \subseteq R_{i:\alpha}$.

The constraint $S3$ says that if w' is the *next* world of w which is reachable from w through the occurrence of agent i's action α, then w' is a world which is *possibly* reachable from w through the occurrence of agent i's action α. The semantic constraint $S3$ corresponds to the following axiom $\textbf{Inc}_{Does,After}$.

$\textbf{Inc}_{Does,After}$ $\text{Does}_{i:\alpha}\varphi \rightarrow \neg\text{After}_{i:\alpha}\neg\varphi$

According to $\textbf{Inc}_{Does,After}$, if i is going to do α and φ will be true afterward, then it is not the case that $\neg\varphi$ will be true after i does α.

The following axioms relate intentions with actions.

IntAct1 $(\text{Int}_i(\alpha) \wedge \text{Can}_i(\alpha)) \rightarrow \text{Does}_{i:\alpha}\top$
IntAct2 $\text{Does}_{i:\alpha}\top \rightarrow \text{Int}_i(\alpha)$

According to **IntAct1**, if i has the intention to do action α and has the capacity to do α, then i is going to do α. According to **IntAct2**, an agent is going to do action α only if he has the intention to do α. In this sense we suppose that an agent's *doing* is by definition intentional. Similar axioms have been studied in [26] in which a logical model of the relationships between intention and action performance is proposed. **IntAct1** and **IntAct2** correspond to the following semantic constraints over $\mathcal{L}$ models. For every $i \in AGT$, $\alpha \in ACT$ and $w \in W$:

S4   if $R_{i:\alpha}$ is defined at w and $D_{i:\alpha}$ is defined at w' for all $w' \in G_i(w)$ then $D_{i:\alpha}$ is defined at w;

S5   if $w' \in G_i(w)$ and $D_{i:\alpha}$ is defined at w, then $D_{i:\alpha}$ is defined at w'.

As far as informative actions are concerned, we assume that they are always executable, i.e. an agent i can always inform another agent j about a fact P. Formally:

CanInf $\text{Can}_i(inf_j(P))$

Note that different from public announcement logic (see, e.g., [17]), in our logic even 'non-truthful' informative actions are executable (i.e. informing that p even if p is currently false).

Axiom **CanInf** corresponds to the following semantic constraint over $\mathcal{L}$ models. For every $i \in AGT$, $inf_j(P) \in INFO$ and $w \in W$:

S6 $R_{i:inf_j(P)}$ is defined at w.

Beliefs, goals and actions. We suppose that goals and beliefs must be compatible, that is, if an agent has the goal that φ then, he cannot believe that $\neg\varphi$. Indeed, the notion of goal we characterize here is a notion of an agent's *chosen goal*, i.e. a goal that an agent decides to pursue. As some authors have stressed (*e.g.*[6]), a rational agent cannot decide to pursue a certain state of affairs φ, if he believes that $\neg\varphi$. Thus, for any $i \in AGT$ and $w \in W$ the following semantic constraint over $\mathcal{L}$ models is supposed:

S7 $G_i(w) \cap B_i(w) \neq \emptyset$.

The constraint $S7$ corresponds to the following axiom **WR** (*weak realism*) of our logic.

WR $\text{Goal}_i\varphi \rightarrow \neg\text{Bel}_i\neg\varphi$

We also assume positive and negative introspection over (chosen) goals, that is:

PIntrGoal $\text{Goal}_i\varphi \rightarrow \text{Bel}_i\text{Goal}_i\varphi$
NIntrGoal $\neg\text{Goal}_i\varphi \rightarrow \text{Bel}_i\neg\text{Goal}_i\varphi$

Axioms **PIntrGoal** and **NIntrGoal** correspond together to the following semantic constraint over $\mathcal{L}$ models. For any $i \in AGT$ and $w \in W$:

S8   if $w' \in B_i(w)$ then $G_i(w) = G_i(w')$.

We suppose that agents satisfy the property of *no forgetting* (**NF**)[4], that is, if an agent i believes that after agent j does α, it is the case that φ, and agent i does not believe that j cannot do action α, then after agent j does α, i believes that φ.

NF $(\text{Bel}_i\text{After}_{j:\alpha}\varphi \wedge \neg\text{Bel}_i\neg\text{Can}_j(\alpha)) \rightarrow \text{After}_{j:\alpha}\text{Bel}_i\varphi$

Axiom **NF** corresponds to the following semantic constraint over $\mathcal{L}$ models. For any $i, j \in AGT$, $\alpha \in ACT$, and $w \in W$:

S9   if there exists v such that $(w, v) \in B_i \circ R_{j:\alpha}$ then $R_{j:\alpha} \circ B_i \subseteq B_i \circ R_{j:\alpha}$.

where $\circ$ is the standard composition operator between two binary relations. In accepting the axiom **NF**, we suppose that events are always uninformative, that is, i should not forget anything about the particular effects of j's action α that starts at a world w. What an agent i believes at a world w', only depends on what i believed at the previous world w and on the action which has occurred and which was responsible for the transition from w to w'. Besides, the axiom **NF** relies on an additional assumption of complete and correct information. It is supposed that j's action α occurs if and only if every agent is informed of this fact. Hence all action occurrences are supposed to be public.

Finally we have specific properties for informative actions. We suppose that if an agent i is informed (resp. not informed) by another j that some fact P is true then i is aware of being informed (resp. not being informed) by j.

PIntrInf $\text{Inf}_{j,i}(P) \rightarrow \text{Bel}_i\text{Inf}_{j,i}(P)$
NIntrInf $\neg\text{Inf}_{j,i}(P) \rightarrow \text{Bel}_i\neg\text{Inf}_{j,i}(P)$

Axioms **PIntrInf** and **NIntrInf** correspond to the following semantic constraints over $\mathcal{L}$ models. For any $i, j \in AGT$, $inf_i(P) \in INFO$, and $w \in W$:

S10   if $D_{j:inf_i(P)}$ is defined at w then for all $v \in B_i(w)$, $D_{j:inf_i(P)}$ is defined at v;
S11   if $D_{j:inf_i(P)}$ is not defined at w then for all $v \in B_i(w)$, $D_{j:inf_i(P)}$ is not defined at v.

---

[4] See, e.g., [14] for a discussion of this property.

We call $\mathcal{L}$ the logic axiomatized by all propositional tautologies, and the axioms and rules of inference **MP**, $\mathbf{K}_{Bel}$, $\mathbf{K}_{Goal}$, $\mathbf{K}_{After}$, $\mathbf{K}_{Does}$, $\mathbf{D}_{Bel}$, $\mathbf{D}_{Goal}$, $\mathbf{4}_{Bel}$, $\mathbf{5}_{Bel}$, $\mathbf{Nec}_{Bel}$, $\mathbf{Nec}_{Goal}$, $\mathbf{Nec}_{After}$, $\mathbf{Nec}_{Does}$, $\mathbf{Alt}_{Does}$, **Active**, $\mathbf{Inc}_{Does,After}$, **IntAct1**, **IntAct2**, **CanInf**, **WR**, **PIntrGoal**, **NIntrGoal**, **NF**, **PIntrInf** and **NIntrInf**.

We write $\vdash \varphi$ if formula φ is a theorem of $\mathcal{L}$ (i.e. φ is the derivable from the axioms and rules of inference of the logic $\mathcal{L}$). We write $\models \varphi$ if φ is *valid* in all $\mathcal{L}$ models, i.e. $M, w \models \varphi$ for every $\mathcal{L}$ model M and world w in M. Finally, we say that φ is *satisfiable* if there exists a $\mathcal{L}$ model M and world w in M such that $M, w \models \varphi$. We can prove that the logic $\mathcal{L}$ is *sound* and *complete* with respect to the class of $\mathcal{L}$ models. Namely:

Theorem 1. $\vdash \varphi$ *if and only if* $\models \varphi$.

Proof. It is a routine task to check that the axioms of the logic $\mathcal{L}$ correspond one-to-one to their semantic counterparts on the frames. In particular, Axioms $\mathbf{D}_{Bel}$, $\mathbf{4}_{Bel}$ and $\mathbf{5}_{Bel}$ correspond to the seriality, transitivity and Euclideanity of every relation B_i. Axiom $\mathbf{D}_{Goal}$ corresponds to the seriality of every relation G_i. Axiom $\mathbf{Alt}_{Does}$ corresponds to the constraint $S1$. Axiom **Active** corresponds to the constraint $S2$. Axiom $\mathbf{Inc}_{Does,After}$ corresponds to the constraint $S3$. Axioms **IntAct1** and **IntAct2** correspond to the constraints $S4$ and $S5$. Axiom **CanInf** corresponds to the constraint $S6$. Axiom **WR** corresponds to the semantic constraint $S7$. Axioms **PIntrGoal** and **NIntrGoal** correspond together to the constraint $S8$. Axiom **NF** corresponds to the constraint $S9$. Axioms **PIntrInf** and **NIntrInf** correspond to the constraints $S10$ and $S11$.

It is routine, too, to check that all of our axioms are in the Sahlqvist class. This means that the axioms are all expressible as first-order conditions on frames and that they are complete with respect to the defined frames classes, cf. [5, Th. 2.42]. □

3   A General Definition of Trust

In this work trust is conceived as a complex configuration of mental states in which there is both a motivational component and an epistemic component. More precisely, we assume that an agent i's trust in agent j necessarily involves a goal of the truster: if agent i trusts agent j then, necessarily, i trusts j with respect to some of his goals. The core of trust is a belief of the truster about some properties of the trustee, that is, if agent i trusts agent j then necessarily i trusts j because i has some goal and believes that j has the right properties to ensure that such a goal will be achieved. The concept of trust formalized in this work is similar to the concept of trust defined by Castelfranchi & Falcone [15,8]. We agree with them that trust should not be seen as an unitary and simplistic notion as other models implicitly suppose. For instance, there are computational models of trust in which trust is conceived as an expectation of the truster about a successful performance of the trustee sustained by the repeated direct interactions with the trustee (under the assumption that iterated experiences of success strengthen the truster's confidence) [23]. More sophisticated models of social trust have been developed in which reputational information is added to information obtained via direct interaction (*e.g.*[21,31]). All these models are in our view over-simplified since they do not consider the beliefs supporting the truster's evaluation of the trustee.

On this point we agree with Castelfranchi & Falcone on the fact that trust is based on the truster's *evaluation* of specific properties of the trustee (*e.g.*abilities, competencies, dispositions, etc.) and of the environment in which the trustee is going to act, which are relevant for the achievement of a goal of the truster. From this perspective, trust is nothing more than the truster's belief about some relevant properties of the trustee with respect to a given goal. [5]

3.1 Trust in the Trustee's Action

The following is the concept of trust as an *evaluation* that interests us in this paper.

Definition 1. *Trust in the Trustee's Action. Agent i trusts j to do α with regard to his goal that φ if and only if i wants φ to be true and i believes that:*[6]

1. *j, by doing α, will ensure that φ AND*
2. *j has the capacity to do α AND*
3. *j intends to do α.*

The formal translation of Definition 1 is:
$$\text{Goal}_i X \varphi \land \text{Bel}_i(\text{After}_{j:\alpha}\varphi \land \text{Can}_j(\alpha) \land \text{Int}_j(\alpha)).$$
In our logic the conditions $\text{Can}_j(\alpha)$ and $\text{Int}_j(\alpha)$ together are equivalent to $\text{Does}_{j:\alpha}\top$ (by Axioms **Inc**$_{Does,After}$, **IntAct1** and **IntAct2**), so the definition of trust in the trustee's action can be simplified as follows:
$$\text{Trust}(i, j, \alpha, \varphi) \overset{\text{def}}{=} \text{Goal}_i X \varphi \land \text{Bel}_i(\text{After}_{j:\alpha}\varphi \land \text{Does}_{j:\alpha}\top).$$
$\text{Trust}(i, j, \alpha, \varphi)$ is meant to stand for: i trusts j to do α with regard to his goal that φ.

Example 1. The two agents i and j are making a transaction in Internet. After having paid j, i trusts j to send him a certain product with regard to his goal of having the product in the next state:
$$\text{Trust}(i, j, send, hasProduct(i)).$$
This means that i wants to have the product in the next state:
$$\text{Goal}_i X\ hasProduct(i).$$
Moreover, according to i's beliefs, j, by sending him the product, will ensure that he will have the product in the next state, and j is going to send the product:
$$\text{Bel}_i(\text{After}_{j:send} hasProduct(i) \land \text{Does}_{j:send}\top).$$
The following theorem highlights the fact that if i trusts j to do α with regard to his goal that φ then i has a positive expectation that φ will be true in the next state.

Theorem 2. *Let $i, j \in AGT$ and $\alpha \in ACT$. Then:*

$$\vdash \text{Trust}(i, j, \alpha, \varphi) \to \text{Bel}_i X\varphi \tag{1}$$

Proof. $\text{Trust}(i, j, \alpha, \varphi)$ implies $\text{Bel}_i(\text{After}_{j:\alpha}\varphi \land \text{Does}_{j:\alpha}\top)$. $\text{After}_{j:\alpha}\varphi \land \text{Does}_{j:\alpha}\top$ implies $\text{Does}_{j:\alpha}\varphi$ (by Axiom **Inc**$_{Does,After}$ and standard principles of the normal operator $\text{Does}_{j:\alpha}$). $\text{Does}_{j:\alpha}\varphi$ implies $X\varphi$. We conclude that $\text{Bel}_i(\text{After}_{j:\alpha}\varphi \land \text{Does}_{j:\alpha}\top)$ implies $\text{Bel}_i X\varphi$ (by Axiom **K**$_{Bel}$). □

---

[5] In this paper we do not consider a related notion of *decision to trust*, that is, the truster's decision to bet and wager on the trustee and to rely on her for the accomplishment of a given task. For a distinction between trust as an *evaluation* and trust as a *decision*, see [15,27].

[6] In the present paper we only focus on *full trust* involving a *certain belief* of the truster. In order to extend the present analysis to forms of *partial trust*, a notion of *graded belief* (i.e. uncertain belief) or *graded trust* is needed.

3.2   Trust in the Trustee's Inaction

In our view *trust in the trustee's action* must be distinguished from *trust in the trustee's inaction*. The former is focused on the domain of gains whereas the latter is focused on the domain of losses. That is, in the former case the truster believes that the trustee is in condition to *further* the achievement (or the maintenance) of a pleasant state of affairs, and it will *do* that; in the latter case the truster believes that the trustee is in condition to *endanger* the achievement (or the maintenance) of a pleasant state of affairs, but it will *refrain* from doing that. Trust in the trustee's inaction is based on the fact that, by doing some action α, agent j can prevent i to reach his goal. In that case i expects that j will not intend to do α.

Definition 2. *Trust in the Trustee's Inaction. Agent i trusts j not to do α with regard to his goal φ if and only if i wants φ to be true and i believes that:*

  1. *j, by doing α, will ensure that $\neg\varphi$ AND*
  2. *j has the capacity to do α AND*
  3. *j does not intend to do α.*

The formal definition of trust in the trustee's inaction is given by the following abbreviation:

$$\text{Trust}(i,j,\sim\alpha,\varphi) \stackrel{def}{=} \text{Goal}_i\text{X}\varphi \wedge \text{Bel}_i(\text{After}_{j:\alpha}\neg\varphi \wedge \text{Can}_j(\alpha) \wedge \neg\text{Int}_j(\alpha)).$$

$\text{Trust}(i,j,\sim\alpha,\varphi)$ stands for: i trusts j not to do α with regard to his goal that φ.

Example 2. Agent j is the webmaster of a public access website with financial information. Agent i is a regular reader of this website and he trusts j not to restrict the access to the website with regard to his goal of having free access to the website:

$$\text{Trust}(i,j,\sim restrict, freeAccess(i)).$$

This means that, i has the goal of having free access to the website in the next state:

$$\text{Goal}_i\text{X} \, freeAccess(i).$$

Moreover, according to i's beliefs, j has the capacity to restrict the access to the website and, by restricting the access to the website, j will ensure that i will not have free access to the website in the next state, but j does not intend to restrict the access:

$$\text{Bel}_i(\text{After}_{j:restrict}\neg freeAccess(i) \wedge \text{Can}_j(restrict) \wedge \neg\text{Int}_j(restrict)).$$

In this situation, i's trust in j is based on i's belief that j is in condition to restrict the access to the website, but she does not have the intention to do this.

Note that, differently from agent i's trust in agent j's action, agent i's trust in agent j's inaction with respect to the goal that φ does not entail i's positive expectation that φ will be true. Indeed, $\text{Trust}(i,j,\sim\alpha,\varphi) \wedge \neg\text{Bel}_i\text{X}\varphi$ is satisfiable in our logic. The intuitive reason is that $\neg\varphi$ may be the effect of another action than $j:\alpha$.

In the following Sections 4 and 5 we will study the properties of information sources and show how these properties can be evaluated by the truster in order to assess the trustworthiness of an information source.

4   Some Basic Properties of an Information Source

We here consider four basic properties of an information source: validity, completeness, sincerity and cooperativity.

Definition 3. *Information Source Validity. Agent j is a valid information source about P with regard to i if and only if, after j does the action of informing i about P, it is the case that P.*

Formally: $\text{Val}(j, i, P) \overset{\text{def}}{=} \text{After}_{j:inf_i(P)} P.$

Definition 4. *Information Source Completeness. Agent j is a complete information source about P with regard to i if and only if, if P is true then j does the action of informing i about P.*

Formally: $\text{Compl}(j, i, P) \overset{\text{def}}{=} P \rightarrow \text{Inf}_{j,i}(P).$

Definition 5. *Information Source Sincerity. Agent j is a sincere information source about P with regard to i if and only if, after j does the action of informing i about P, it is the case that j believes P.*

Formally: $\text{Sinc}(j, i, P) \overset{\text{def}}{=} \text{After}_{j:inf_i(P)} \text{Bel}_j P.$

Definition 6. *Information Source Cooperativity. Agent j is a cooperative information source about P with regard to i if and only if, if j believes that P then j informs i about P.*

Formally: $\text{Coop}(j, i, P) \overset{\text{def}}{=} \text{Bel}_j P \rightarrow \text{Inf}_{j,i}(P).$

Example 3. Consider an example in the field of stocks and bonds market. The agent BUG is the Bank of Union of Groenland. Sue Naive (SN) and Very Wise (VW) are two BUG's customers. BUG plays the role of an information source for the customers, for instance for the facts p: "it is recommended to buy MicroHard stocks", and q: "Microhard stocks are dropping". SN believes that BUG is sincere with regard to her about p, because SN believes that BUG wants to help its customers and BUG has a long experience in the domain. SN also believes that BUG is cooperative with regard to her about q because q is a relevant information for customers in order to make decisions. VW does not believe that BUG is sincere with regard to him about p. Indeed, VW believes that BUG wants that VW buys Microhard stocks, even if this is not profitable for VW. This example is formally represented by the following formula: $\text{Bel}_{SN}(\text{Sinc}(BUG, SN, p) \wedge \text{Coop}(BUG, SN, q)) \wedge \neg\text{Bel}_{VW}\text{Sinc}(BUG, VW, p).$

5   Trust in Information Sources

We conceive trust in information sources as a specific instance of the general notion of trust in the trustee's action defined in Section 3. In our view, the relevant aspect of trust in information sources is the content of the truster's goal. In particular, we suppose that an agent i trusts the information source j to inform him whether the fact P is true only if i has the *epistemic goal* of knowing whether P is true and believes that, due to the information transmitted by j, he will achieve this goal. In this sense, trust in information sources is characterized by an epistemic goal of the truster and an informative action of the trustee. The concept of epistemic goal can be defined from the following standard definitions of *knowing that* (i.e. as having the correct belief that something is the case) and *knowing whether*:

$$\mathtt{K}_i\varphi \overset{\mathrm{def}}{=} \mathtt{Bel}_i\varphi \wedge \varphi \quad \mathtt{Kw}_i\varphi \overset{\mathrm{def}}{=} \mathtt{K}_i\varphi \vee \mathtt{K}_i\neg\varphi$$

where $\mathtt{K}_i\varphi$ stands for "agent i knows that φ is true" and, $\mathtt{Kw}_i\varphi$ stands for "i knows whether φ is true". An *epistemic goal* of an agent i is i's goal of knowing the truth value of a certain formula. Formally, $\mathtt{Goal}_i\mathtt{Kw}_i\varphi$ denotes i's epistemic goal of knowing whether φ is true now; $\mathtt{Goal}_i\mathtt{X}\,\mathtt{Kw}_i\varphi$ denotes i's epistemic goal of knowing whether φ is true in the next state.

Our aim in this section of the paper is to investigate the relationships between trust in information sources and the properties of information sources defined in Section 4. The following Theorem 3 highlights the relationship between trust in information sources and the properties of validity and completeness of information sources. It says that: if i believes that j is a valid information source about p and $\neg p$ with regard to i and that j is a complete information source about p and $\neg p$ with regard to i and, i has the epistemic goal of knowing whether p is true then, either i trusts the information source j to inform him that p is true or i trusts the information source j to inform him that $\neg p$ is true (with regard to his epistemic goal of knowing whether p is true).

Theorem 3. *Let $i, j \in AGT$ and $inf_i(p), inf_i(\neg p) \in INFO$, then:*

$$\vdash (\mathtt{Bel}_i(\mathtt{Val}(j,i,p) \wedge \mathtt{Val}(j,i,\neg p)) \wedge \mathtt{Bel}_i(\mathtt{Compl}(j,i,p) \wedge \mathtt{Compl}(j,i,\neg p)) \wedge \tag{2}$$
$$\mathtt{Goal}_i\mathtt{X}\,\mathtt{Kw}_ip) \rightarrow (\mathtt{Trust}(i,j,inf_i(p),\mathtt{Kw}_ip) \vee \mathtt{Trust}(i,j,inf_i(\neg p),\mathtt{Kw}_ip))$$

Proof. We give a sketch of the proof. By Axioms $\mathbf{K}_{Bel}$, **PIntrInf** and **NIntrInf**, $\mathtt{Bel}_i(\mathtt{Compl}(j,i,p) \wedge \mathtt{Compl}(j,i,\neg p))$ implies $\mathtt{Bel}_i\mathtt{Inf}_{j,i}(p) \vee \mathtt{Bel}_i\mathtt{Inf}_{j,i}(\neg p)$.

By Axioms $\mathbf{K}_{Bel}$, **Nec**$_{Bel}$, **CanInf**, $\mathbf{D}_{Bel}$, $\mathbf{4}_{Bel}$, $\mathbf{5}_{Bel}$ and **NF**, $\mathtt{Bel}_i\mathtt{Inf}_{j,i}(p) \vee \mathtt{Bel}_i\mathtt{Inf}_{j,i}(\neg p)$ and $\mathtt{Bel}_i(\mathtt{Val}(j,i,p) \wedge \mathtt{Val}(j,i,\neg p))$ and $\mathtt{Goal}_i\mathtt{X}\,\mathtt{Kw}_ip$ imply $(\mathtt{Goal}_i\mathtt{X}\,\mathtt{Kw}_ip \wedge \mathtt{Bel}_i\mathtt{After}_{j:inf_i(p)}\mathtt{Kw}_ip \wedge \mathtt{Bel}_i\mathtt{Inf}_{j,i}(p)) \vee (\mathtt{Goal}_i\mathtt{X}\,\mathtt{Kw}_ip \wedge \mathtt{Bel}_i\,\mathtt{After}_{j:inf_i(\neg p)}\mathtt{Kw}_ip \wedge \mathtt{Bel}_i\mathtt{Inf}_{j,i}(\neg p))$. The latter is equivalent to $\mathtt{Trust}(i,j,inf_i(p),\mathtt{Kw}_ip) \vee \mathtt{Trust}(i,j,inf_i(\neg p),\mathtt{Kw}_ip)$. $\square$

Example 4. Let us consider again the example of stocks and bonds market. SN has the epistemic goal of knowing whether q ("Microhard stocks are dropping") is true:

$$\mathtt{Goal}_{SN}\mathtt{X}\,\mathtt{Kw}_{SN}q.$$

SN believes that BUG is a valid information source with regard to her both about q and about $\neg q$ and that BUG is a complete information source with regard to her both about q and about $\neg q$:

$$\mathtt{Bel}_{SN}(\mathtt{Val}(BUG,SN,q) \wedge \mathtt{Val}(BUG,SN,\neg q)) \wedge$$
$$\mathtt{Bel}_{SN}(\mathtt{Compl}(BUG,SN,q) \wedge \mathtt{Compl}(BUG,SN,\neg q)).$$

Then, by Theorem 3, we can infer that either SN trusts the information source BUG to inform her that q is true or SN trusts the information source BUG to inform her that $\neg q$ is true (with regard to her epistemic goal of knowing whether q is true):

$\texttt{Trust}(SN, BUG, inf_{SN}(q), \texttt{Kw}_{SN}q) \lor \texttt{Trust}(SN, BUG, inf_{SN}(\neg q), \texttt{Kw}_{SN}q)$.
Finally, by Theorem 2, we can infer that SN believes that in the next state she will achieve her goal of knowing whether q is true:

$$\texttt{Bel}_{SN}\texttt{X Kw}_{SN}q.$$

In the following two Sections 6 and 7 we will shift the focus of analysis from information sources to communication systems. We will study some important properties of communication systems and show how these properties can be evaluated by the truster in order to assess the trustworthiness of a communication system.

6 Some Basic Properties of a Communication System

We suppose that the fundamental properties of a communication system j can be defined in terms of two facts: the communication system j satisfies an agent i's goal that a certain information will be transmitted to another agent z or, the communication system j satisfies an agent i's goal that a certain information will not be transmitted to another agent z. In the former case we say that the communication system j is available to i to transmit the information. In the latter case we say that the communication system j ensures to i the privacy of the information. For simplification, we ignore in this work other properties of communication systems like authentication and integrity.

Definition 7. *Communication System Availability. The communication system j is available to agent i to transmit the information P to agent z if and only if, if j believes that i wants that j informs z about P then j informs z about P.*

Formally: $\texttt{Avail}(j, i, z, P) \overset{\text{def}}{=} \texttt{Bel}_j\texttt{Goal}_i\texttt{Inf}_{j,z}(P) \to \texttt{Inf}_{j,z}(P)$.

Definition 8. *Communication System Privacy. The communication system j ensures to agent i the privacy of information P from agent z if and only if, if j believes that i wants that j does not inform z about P then, j does not inform z about P.*

Formally: $\texttt{Priv}(j, i, z, P) \overset{\text{def}}{=} \texttt{Bel}_j\texttt{Goal}_i\neg\texttt{Inf}_{j,z}(P) \to \neg\texttt{Inf}_{j,z}(P)$.

Example 5. Let us consider an example in the field of Web services. An agent called Bill decides to use a Hotel Booking Service (HBS) in Internet in order to book a double room at the Hotel Colosseum (HC) in Rome. Bill's decision is affected by two beliefs of Bill, the belief that HBS ensures the privacy from a potential intruder of the information r: "Bill's credit card number is 01234567891234", the belief that HBS is available to inform HC about s: "Bill has made an online reservation". According to our definitions, this example is formally represented by:

$$\texttt{Bel}_{Bill}\texttt{Priv}(HBS, Bill, intruder, r) \land \texttt{Bel}_{Bill}\texttt{Avail}(HBS, Bill, HC, s).$$

7 Trust in Communication Systems

We conceive trust in communication systems as a specific instance of the notion of trust defined in Section 3. On the one hand, we suppose that an agent i trusts the communication system j to inform agent z about P with regard to his goal that z believes P

(*trust in a communication system's action*) if and only if, i has the goal that z believes P and i believes that, due to the information transmitted by j to z, z will believe P. On the other hand, we suppose that an agent i trusts the communication system j not to inform agent z about P with regard to his goal that z does not believe P (*trust in a communication system's inaction*) if and only if, i has the goal that z does not believe P, i believes that, by informing z about P, j will ensure that z believes P but, i believes that j does not intend to inform z about P. In this sense, i's trust in the communication system j's action (resp. inaction) is characterized by i's goal that a certain information will be transmitted (resp. will not be transmitted) to another agent z so that z will have access (resp. will not have access) to this information.

Our aim in this section of the paper is to investigate the relationships between trust in communication systems and the two properties of communication systems defined in Section 6. The following Theorem 4 highlights the relationship between trust in a communication system's action and the availability of the communication system. It says that: if i has the goal that in the next state z will believe P, i believes that j is available to inform z about P, i believes that j believes that i wants j to inform z about P and, i believes that z believes that j is a valid information source about P then, i trusts j to inform z about P with regard to his goal that z will believe P.

Theorem 4. *Let* $i, j, z \in AGT$ *and* $inf_z(P) \in INFO$, *then:*

$$\vdash (\texttt{Goal}_i\texttt{X Bel}_z P \land \texttt{Bel}_i\texttt{Avail}(j, i, z, P) \land \texttt{Bel}_i\texttt{Bel}_j\texttt{Goal}_i\texttt{Inf}_{j,z}(P) \land \\ \texttt{Bel}_i\texttt{Bel}_z\texttt{Val}(j, z, P)) \rightarrow \texttt{Trust}(i, j, inf_z(P), \texttt{Bel}_z P) \tag{3}$$

Proof. We give a sketch of the proof. By $\mathbf{K}_{Bel}$, $\mathbf{Nec}_{Bel}$, $\mathbf{D}_{Bel}$, **CanInf**, and **NF**, $\texttt{Bel}_i\texttt{Avail}(j, i, z, P) \land \texttt{Bel}_i\texttt{Bel}_j\texttt{Goal}_i\texttt{Inf}_{j,z}(P) \land \texttt{Bel}_i\texttt{Bel}_z\texttt{Val}(j, z, P)$ implies $\texttt{Bel}_i\texttt{Inf}_{j,z}(P) \land \texttt{Bel}_i\texttt{After}_{j:inf_z(P)}\texttt{Bel}_z P$.

The latter is equivalent to $\texttt{Trust}(i, j, inf_z(P), \texttt{Bel}_z P)$. $\square$

Example 6. Let us consider again the example of the Hotel Booking Service (HBS). Bill has the goal that the receptionist at the Hotel Colosseum (HC) believes that s ("Bill has made an online reservation"):

$$\texttt{Goal}_{Bill}\texttt{X Bel}_{HC}s.$$

Bill believes that HBS is available to inform the HC's receptionist about s and that HBS believes that Bill wants HBS to inform the HC's receptionist about s:

$$\texttt{Bel}_{Bill}\texttt{Avail}(HBS, Bill, HC, s) \land \texttt{Bel}_{Bill}\texttt{Bel}_{HBS}\texttt{Goal}_{Bill}\texttt{Inf}_{HBS,HC}(s).$$

Bill also believes that the HC's receptionist believes that HBS is a valid information source about s:

$$\texttt{Bel}_{Bill}\texttt{Bel}_{HC}\texttt{Val}(HBS, HC, s).$$

Then, from Theorem 4, we can infer that Bill trusts HBS to inform the HC's receptionist about s with regard to his goal that the HC's receptionist will believe s:

$$\texttt{Trust}(Bill, HBS, inf_{HC}(s), \texttt{Bel}_{HC}s).$$

The following Theorem 5 highlights the relationship between trust in a communication system's inaction and the fact that the communication system ensures privacy. It says that: if i has the goal that z does not believe P, i believes that j ensures the privacy of information P from agent z, i believes that j believes that i wants that j does not inform

z about P and, i believes that z believes that j is a valid information source about P then, i trusts j not to inform z about P with regard to his goal that z does not believe P.

Theorem 5. *Let* $i, j, z \in AGT$ *and* $inf_z(P) \in INFO$, *then:*

$$\vdash (\mathtt{Goal}_i \mathtt{X} \neg \mathtt{Bel}_z P \wedge \mathtt{Bel}_i \mathtt{Priv}(j, i, z, P) \wedge \mathtt{Bel}_i \mathtt{Bel}_j \mathtt{Goal}_i \neg \mathtt{Inf}_{j,z}(P) \wedge$$
$$\mathtt{Bel}_i \mathtt{Bel}_z \mathtt{Val}(j, z, P)) \rightarrow \mathtt{Trust}(i, j, \sim inf_z(P), \neg \mathtt{Bel}_z P) \tag{4}$$

Proof. We give a sketch of the proof. By $\mathbf{K}_{Bel}$, $\mathbf{Nec}_{Bel}$, $\mathbf{D}_{Bel}$, $\mathbf{CanInf}$ and $\mathbf{NF}$, $\mathtt{Bel}_i \mathtt{Priv}(j, i, z, P) \wedge \mathtt{Bel}_i \mathtt{Bel}_j \mathtt{Goal}_i \neg \mathtt{Inf}_{j,z}(P) \wedge \mathtt{Bel}_i \mathtt{Bel}_z \mathtt{Val}(j, z, P)$ implies $\mathtt{Bel}_i \neg \mathtt{Int}_j(inf_z(P)) \wedge \mathtt{Bel}_i \mathtt{Can}_j(inf_z(P)) \wedge \mathtt{Bel}_i \mathtt{After}_{j:inf_z(P)} \mathtt{Bel}_z P$.
The latter is equivalent to $\mathtt{Trust}(i, j, \sim inf_z(P), \neg \mathtt{Bel}_z P)$. $\qquad\square$

Example 7. In this version of the scenario of the Hotel Booking Service (HBS) Bill has the goal that a potential intruder will not have access to the information r ("Bill's credit card number is 01234567891234"):

$$\mathtt{Goal}_{Bill} \mathtt{X} \neg \mathtt{Bel}_{intruder} r.$$

Bill believes that HBS ensures the privacy of information r from potential intruders and that HBS believes that Bill wants that HBS does not inform a potential intruder about r:

$$\mathtt{Bel}_{Bill} \mathtt{Priv}(HBS, Bill, intruder, r) \wedge \mathtt{Bel}_{Bill} \mathtt{Bel}_{HBS} \mathtt{Goal}_{Bill} \neg \mathtt{Inf}_{HBS, intruder}(r).$$

Finally, Bill believes that every potential intruder believes that HBS is a valid information source about credit card numbers:

$$\mathtt{Bel}_{Bill} \mathtt{Bel}_{intruder} \mathtt{Val}(HBS, intruder, r).$$

Then, from Theorem 5, we can infer that Bill trusts HBS not to inform a potential intruder about r with regard to his goal that a potential intruder will not believe r:

$$\mathtt{Trust}(Bill, HBS, \sim inf_{intruder}(r), \neg \mathtt{Bel}_{intruder} r).$$

8   Related Works

Several logical models of trust in information sources have been proposed in the recent literature [24,12,11]. Some of them take the concept of trust as a primitive [24], whereas others reduce trust to a kind of belief of the truster [12]. All these logical models do not investigate the epistemic supports for this form of trust. Moreover, they do not consider the motivational aspect of trust. In the present paper both aspects have been taken into account. On the one hand, we have modeled the properties of information sources such as sincerity, validity, completeness and cooperativity and shown that some of them are epistemic supports for an agent's trust in an information source, that is, they are sufficient conditions for trusting an information source to inform whether a certain fact is true (Theorem 3). On the other hand, we have modeled the truster's epistemic goal of knowing whether a certain fact is true.

More generally, the truster's goal is for us a necessary constituent of the definition of trust (see [13,15] for a similar view of trust). In the definitions of trust in the trustee's action and trust in the trustee's inaction proposed in Section 3, i's trust in j necessarily involves a main and primary motivational component which is a goal of the truster. If i trusts j then, necessarily i trusts j because i has some goal and thinks that j has the right properties to ensure that such a goal will be achieved. The goal component in the definition of trust is for us fundamental since it allows to distinguish trust from mere *thinking* and *foreseeing*. In this sense, our logical approach to trust is different from the approach proposed by Jones [22] in which the motivational aspect is not considered to be necessary for defining trust, and trust is characterized only in terms of two beliefs of the truster: the truster's belief that a certain rule or regularity applies to the trustee (called "rule belief"), and the truster's belief that the rule or regularity is going to be followed by the trustee (called "conformity belief").

As far as communication systems are concerned, there are several logical models in the literature which deal with the security properties of a communication system such as privacy (or confidentiality), availability, integrity, authentication (see *e.g.*[10,7]). Nevertheless, there is still no formal analysis of the relationships between these security properties of a communication system and trust in the communication system. In this work such relationships have been clarified (Theorems 4 and 5). It is also to be noted that the definitions of privacy and availability given in Section 6 are based on an agent i's goal that another agent j will inform (resp. will not inform) a third agent z about a certain fact P. Thus, our definitions are different from the definitions proposed in the security domain in which privacy and availability are traditionally expressed in terms of normative concepts like authorization and permission. According to [4] for instance privacy (or confidentiality) means that "(...) data may only be learned by authorized users". According to [18], privacy (or confidentiality) "(...) is the concept of: 1) ensuring that information is accessible for reading, listening, recording or physical removal only to subjects entitled to it, and 2) that subjects only read or listens to the information to the extent permitted". A variant of confidentiality based on normative concepts and closer to the definition of confidentiality proposed in the security domain could be expressed in our logical framework. To this aim, it would be sufficient to substitute the goal modality $\texttt{Goal}_i$ with an obligation modality $\texttt{Obl}$ of Standard Deontic Logic [1] in the definition given in Section 6. That is:

$$\texttt{Priv}(j, z, P) \overset{\text{def}}{=} \texttt{Bel}_j \texttt{Obl} \neg \texttt{Inf}_{j,z}(P) \rightarrow \neg \texttt{Inf}_{j,z}(P).$$

According to this 3-argument definition, the communication system j ensures the privacy of information P from agent z if and only if, if j believes that it is obligatory that he does not inform z about P then, j does not inform z about P. This definition can be expressed in an equivalent form by the notion permission, the dual of obligation:

$$\texttt{Priv}(j, z, P) \leftrightarrow (\texttt{Bel}_j \neg \texttt{Perm}\, \texttt{Inf}_{j,z}(P) \rightarrow \neg \texttt{Inf}_{j,z}(P)).$$

That is, the communication system j ensures the privacy of information P from agent z if and only if, if j believes that it is not permitted that he informs z about P then, j does not inform z about P.

9   Conclusion

We have presented in a modal logical framework a model that integrates in the definition of trust: the truster's goal, the trustee's action that ensures the achievement of the truster's goal, and the trustee's ability and intention to do this action. In the same logical framework we have defined several properties of information sources (validity, completeness, sincerity and cooperativity) and discussed their relationships with an agent's trust in an information source. In the last part of the paper, we have investigated some properties of communication systems (availability and privacy) and discussed their relationships with an agent's trust in a communication system. It is to be noted that, due to the complexity of the concepts involved in our analysis of trust, we had to accept strong simplifications. For instance, in the definitions of the properties of information sources and communication systems entailment is formalized by a material implication, while some form of conditional might be more adequate. Moreover, in the future we intend to provide a more fine-grained semantics of the notion of informative action. We think that formalizing informative actions as *private announcements* in the sense of [2] would be an interesting improvement of the logical theory of trust presented in this paper.

Acknowledgements

This research has been supported by the project ForTrust "Social trust analysis and formalization" ANR-06-SETI-006 financed by the French institution "Agence Nationale de la Recherche".

References

1. Åqvist, L.: Deontic logic. In: Gabbay, D.M., Geunther, F. (eds.) Handbook of Philosophical Logic. Kluwer Academic Publishers, Dordrecht (2002)
2. Baltag, A., Moss, L., Solecki, S.: The logic of public announcements, common knowledge and private suspicions. In: Proceedings of the Seventh Conference on Theoretical Aspects of Rationality and Knowledge (TARK 1998), pp. 43–56. Morgan Kaufmann, San Francisco (1998)
3. Berners-Lee, T., Hendler, J., Lassila, O.: The semantic web. Scientific American 284(5), 34–43 (2001)
4. Bieber, P., Cuppens, F.: Expression of confidentiality policies with deontic logic. In: Deontic logic in computer science: normative system specification, pp. 103–123. John Wiley and Sons, Chichester (1993)
5. Blackburn, P., de Rijke, M., Venema, Y.: Modal Logic. Cambridge University Press, Cambridge (2001)
6. Bratman, M.: Intentions, plans, and practical reason. Harvard University Press (1987)
7. Burrows, M., Abadi, M., Needham, R.M.: A logic of authentication. ACM Transactions on Computer Systems 8(1), 18–36 (1990)
8. Castelfranchi, C., Falcone, R.: Principles of trust for MAS: Cognitive anatomy, social importance, and quantification. In: Proceedings of the Third International Conference on Multiagent Systems (ICMAS 1998), pp. 72–79 (1998)
9. Cohen, P.R., Levesque, H.J.: Intention is choice with commitment. Artificial Intelligence 42, 213–261 (1990)
10. Cuppens, F., Demolombe, R.: A Deontic Logic for Reasoning about Confidentiality. In: Proceedings of Third International Workshop on Deontic Logic in Computer Science (DEON 1996), Workshops in Computing, pp. 66–79. Springer, Heidelberg (1996)
11. Dastani, M., Herzig, A., Hulstijn, J., van der Torre, L.: Inferring trust. In: Leite, J., Torroni, P. (eds.) CLIMA 2004. LNCS (LNAI), vol. 3487, pp. 144–160. Springer, Heidelberg (2005)
12. Demolombe, R.: To trust information sources: a proposal for a modal logical framework. In: Castelfranchi, C., Tan, Y.-H. (eds.) Trust and Deception in Virtual Societies. Kluwer, Dordrecht (2001)

13. Deutsch, M.: Trust and suspicion. The Journal of Conflict Resolution 2(4), 265–279 (1958)
14. Fagin, R., Halpern, J., Moses, Y., Vardi, M.: Reasoning about Knowledge. MIT Press, Cambridge (1995)
15. Falcone, R., Castelfranchi, C.: Social trust: A cognitive approach. In: Castelfranchi, C., Tan, Y.H. (eds.) Trust and Deception in Virtual Societies, pp. 55–90. Kluwer, Dordrecht (2001)
16. Fasli, M.: Agent Technology for E-commerce. Wiley & Sons, Chichester (2007)
17. Gerbrandy, J., Groeneveld, W.: Reasoning about information change. Journal of Logic, Language, and Information 6, 147–196 (1997)
18. Hammer, J.H., Schneider, G.: On the Definition and Policies of Confidentiality. In: Proceedings of the Third International Symposium on Information Assurance and Security (IAS 2007), pp. 337–342. IEEE Computer Society, Los Alamitos (2007)
19. Harel, D., Kozen, D., Tiuryn, J.: Dynamic Logic. MIT Press, Cambridge (2000)
20. Herzig, A., Lorini, E., Hubner, J.F., Ben-Naim, J., Castelfranchi, C., Demolombe, R., Longin, D., Vercouter, L.: Prolegomena for a logic of trust and reputation. In: Proceedings of the Third International Workshop on Normative Multiagent Systems, NorMAS 2008 (2008)
21. Huynh, T.G., Jennings, N.R., Shadbolt, N.R.: An integrated trust and reputation model for open multi-agent systems. Journal of Autonomous Agent and Multi-Agent Systems 13, 119–154 (2006)
22. Jones, A.J.I.: On the concept of trust. Decision Support Systems 33(3), 225–232 (2002)
23. Jonker, C.M., Treur, J.: Formal analysis of models for the dynamics of trust based on experiences. In: Garijo, F.J., Boman, M. (eds.) MAAMAW 1999. LNCS, vol. 1647, pp. 221–231. Springer, Heidelberg (1999)
24. Liau, C.J.: Belief, information acquisition, and trust in multi-agent systems: a modal logic formulation. Artificial Intelligence 149, 31–60 (2003)
25. Lorini, E., Demolombe, R.: Trust and norms in the context of computer security. In: van der Meyden, R., van der Torre, L. (eds.) DEON 2008. LNCS (LNAI), vol. 5076, pp. 50–64. Springer, Heidelberg (2008)
26. Lorini, E., Herzig, A.: A logic of intention and attempt. Synthese 163(1), 45–77 (2008)
27. Marsh, S.: Formalising Trust as a Computational Concept. PhD thesis, University of Stirling (1994)
28. McIlraith, S.A., Son, T.C., Zeng, H.: Semantic web services. IEEE Intelligent Systems 16(2), 46–53 (2001)
29. Meyer, J.-J.Ch., van der Hoek, W., van Linder, B.: A logical approach to the dynamics of commitments. Artificial Intelligence 113(1-2), 1–40 (1999)
30. Rao, A.S., Georgeff, M.P.: Modelling rational agents within a BDI-architecture. In: Proceedings of the 2nd International Conference on Principles of Knowledge Representation and Reasoning (KR 1991), pp. 473–484. Morgan Kaufmann, San Francisco (1991)
31. Sabater, J., Sierra, C.: Regret: a reputation model for gregarious societies. In: Proceedings of the First International Joint Conference on Autonomous Agents and Multi-Agent Systems (AAMAS 2002), pp. 475–482. ACM Press, New York (2001)

Pre-processing Techniques for Anytime Coalition Structure Generation Algorithms

Tomasz Michalak, Andrew Dowell, Peter McBurney, and Michael Wooldridge

Department of Computer Science,
The University of Liverpool, UK
{tomasz,adowell,mcburney,mjw}@liv.ac.uk

Abstract. This paper is concerned with optimal coalition structure generation in multi-agent systems. For characteristic function game representations, we propose pre-processing techniques, presented in the form of filter rules, that reduce the intractability of the coalition structure generation problem by identifying coalitions which cannot belong to any optimal structure. These filter rules can be incorporated into many potential anytime coalition structure generation algorithms but we test the effectiveness of these filter rules in the sequential application of the distributed coalition value calculation algorithm (DCVC) [1] and the anytime coalition structure generation algorithm of Rahwan *et al.* (RCSG) [2]. The distributed DCVC algorithm provides an input to the centralised RCSG algorithm and we show that, for both normal and uniform distributions of coalition values, the proposed filter rules reduce the size of this input by a considerable amount. For example, in a system of 20 agents, fewer than 5% of all coalition values have to be input, compared to more than 90% when filter rules are not employed. Furthermore, for a normal distribution of coalition values, the running time of the RCSG algorithm exponentially accelerates as a consequence of the significantly reduced input size. This pre-processing technique bridges the gap between the distributed DCVC and centralised RCSG algorithms and is a natural benchmark to develop a distributed CSG algorithm.[1]

1 Background and Problem Definition

In multi-agent systems (MAS), coalition formation occurs when distinct autonomous agents group together to achieve something more efficiently than they could accomplish individually. Typically, coalition formation is often studied using *characteristic function games* (CFG). These representations consist of a set of agents A and a characteristic function v, which assigns a numerical value to every feasible coalition $C \subseteq A$, reflecting the effectiveness of the co-operation within each coalition of agents . In this convenient but simplified representation, it is assumed that the performance of any one coalition is independent from other co-existing coalitions in the system. In other words, every coalition C has the same value in every structure CS to which it belongs. Evidently, the CFG representation is a special case of the more general *partition function*

---

[1] The authors are grateful for financial support received from the UK EPSRC through the project *Market-Based Control of Complex Computational Systems* (GR/T10657/01).

J.-J.Ch. Meyer and J.M. Broersen (Eds.): KRAMAS 2008, LNAI 5605, pp. 99–113, 2009.

game (PFG) representation in which the value of any coalition depends on other co-operational arrangements between the agents in the whole system [3].

One of the main challenges in co-operative MAS is to determine which exhaustive division of agents into disjoint coalitions (referred to as a *coalition structure* (CS) from now on) maximizes the total payoff to the system. This complex research issue is referred to as the *coalition structure generation* (CSG) problem and since the number of structures increases exponentially as the number of agents increases linearly (for example, there are $190,899,322$ possible structures for 14 agents compared to 1.3 billion structures for 15 agents), the problem of computing an optimal CS becomes time consuming, even for a moderate number of agents. Consequently, much of the research into the CSG problem has focused on generating an optimal CS by evaluating as few coalition structures as possible. This line of research can be divided into three broad categories: (i) limiting the size of coalitions that can be formed [4], (ii) reducing the number of structures that need to be searched at the expense of accuracy [5,6], and (iii) proposing algorithms which take the advantage of the CFG representation to vastly reduce the time taken to generate an optimal CS. Since we are interested in generating optimal solutions in an unconstrained domain, in this paper, we will focus exclusively on the third approach.

There are two general classes of CSG algorithms. Yen [7] proposed an algorithm based on dynamic programming (DP) methods. The advantage of this approach is that it outputs an optimal structure without comparing any of them. However, one disadvantage is that it only outputs an optimal structure after it has completed its entire execution, meaning such methods are not appropriate when the time required to return an optimal solution is longer than the time available to the agents. To circumvent these problems, Rahwan *et al.* [2] proposed an anytime CSG algorithm which divides the space of all structures (denoted Π from now on) into sub-spaces consisting of coalition structures which are identical w.r.t. the sizes of the coalitions involved. Using this representation, and taking, as input, all feasible coalition values, this algorithm uses statistical information, computed from the input coalition values, to determine which of the sub-spaces are 'promising', *i.e.*, which of them may contain an optimal structure. The algorithm then searches these 'promising' subspaces, once again, using statistical information to avoid generating structures which cannot be optimal. This methodology exploits the fact that in CFGs, every coalition C has the same value in every structure CS to which it belongs. Since it is possible to utilize statistical data computed from coalition values to reason about the values of coalition structures, in this paper, we propose a number of pre-processing techniques. These techniques are represented in the form of filter rules which have syntax: *condition* $\rightarrow$ *action*, with the interpretation being that all coalition values which meet the requirements of the condition cannot belong to an optimal structure and so an appropriate action is performed.

Typically, such actions involve filtering coalition values from the input, or filtering all structures containing these coalitions from the search-space. Filtering coalition values from the input is important for two reasons. Firstly, it reduces the number of coalition values an individual agent needs to transfer if the coalition value calculation process is distributed - as in the DCVC algorithm of Rahwan and Jennings [1]. Secondly, it automatically reduces the search space as fewer coalition structures can be created from the input. To test the effectiveness of our approach, we compare the sequential operation of the DCVC and Rahwan *et al.* anytime algorithm both with and without the filter rules. Following the MAS literature, we focus on normal and uniform distributions of coalition values [2] and show that our filter rules:

- always significantly reduce the size of input (from 90% to 5% and 3% for normal and uniform distributions, respectively); and,
- exponentially reduce the time needed to search promising subspaces for a normal distribution whereas they do not affect the performance of the anytime CSG algorithm for a uniform distribution.

2   A Pre-processing Approach to Solve the CSG Problem

Let $A = \{1, \ldots, n\}$ be the set of all agents in the system. Since more than one structure can maximize the value of a system, the output of the CSG process may consist of a set of optimal coalition structures, denoted by $\{CS^*\}$. As of now, the CSG literature for CFG representations has exclusively focused on finding a single optimal coalition structure, denoted by $CS^* \subseteq \{CS^*\}$ [7,2]. Usually, the choice of $CS^* \subseteq \{CS^*\}$ for $|\{CS^*\}| \geq 2$ is made in an *ad hoc* manner, *e.g.*, the optimal coalition structure output is the first CS with maximal value which the algorithm encounters. However, there may be other factors which, although not displayed in the characteristic function, can be used to determine if one structure is better than another and so we consider the $CSG(\{CS^*\})$ problem from now onward. We will denote the set of all feasible coalitions that can be created in the system by F and the corresponding set of all coalition values by $V(F)$.

The most important property of CFG representation is that, as opposed to the general PFG representation, the value of any coalition is independent from the formation of other distinct coalitions in the system. In other words, in a system of $n = |A|$ agents, for any coalition $C \subseteq A$, the value of this coalition $v(C)$ is the same in every possible structure CS where $C \in CS$. Thus, if it can be shown that the value $v(C)$ is too small for C to be in any optimal structure of the system then, clearly, any structure containing C cannot be in the optimal set and can be disregarded. Similarly, if it is proven that the combined value of a group of disjoint coalitions it too small for this group to be in any optimal structure then any structure simultaneously containing every single one of these coalitions can be disregarded. The above discussion can be summarized in the following lemma which holds for CFG representations:

Lemma 1. *For any non-trivial[2] coalition C that can be divided into k disjoint sub-coalitions $C_1, \ldots, C_k$ where $C_1 \cup C_2 \ldots \cup C_k = C$; if $v(C_1) + \ldots + v(C_k) > v(C)$ then $C \notin CS^*$ and so $\forall CS : C \in CS, CS \notin \{CS^*\}$.*

Proof. Consider any coalition structure CS containing coalition C. If CS is an optimal structure (belongs to $\{CS^*\}$) then no other structure can have a value greater than the

---

[2] All coalitions of more than two agents will sometimes be referred to as *non-trivial coalitions*. In contrast, singletons, *i.e.*, agents acting on their own, will sometimes be referred to as *trivial coalitions*.

value of CS. However, if C can be divided into k sub-coalitions $C_1, \ldots, C_k$ where $C_1 \cup C_2 \ldots \cup C_k = C$ and $|C_i| \geq 1 \; \forall i = 1, \ldots, k$ such that $v(C_1) + \ldots v(C_k) > v(C)$ then clearly the structure $CS' = CS \setminus \{C\} \cup \{C_1 \cup \ldots \cup C_k\}$ has a value greater than the value of CS. Therefore CS, where CS is any structure containing coalition C, cannot be an optimal structure and $\forall CS : C \in CS, CS$ does not belong to $\{CS^*\}$.

For example, in a system of five agents $A = \{a_1, a_2, a_3, a_4, a_5\}$ if the value of the coalition $\{a_1, a_2\}$ is less than the sum of the values of the individual coalitions $\{a_1\}$ and $\{a_2\}$ then it is clear that any structure $CS_a = \{a_1, a_2\} \cup CS'$, where CS' is a specific structure of the agents $\{a_3, a_4, a_5\}$, must have value less than the structure $CS_b = \{a_1\}, \{a_2\} \cup CS'$. Consequently, any structure CS which contains coalition $\{a_1, a_2\}$ cannot be optimal and need not be considered when generating an optimal CS.

Existing CSG algorithms take advantage of this characteristic. For example, in the anytime CSG algorithm of Rahwan *et al.* described in Section 1 some subspaces of coalition structures are pruned away from the search space before the search process has begun. Using statistical information obtained from the input of coalition values, it is decided *a priori* if certain coalition structures cannot be optimal. Similarly, in the process of searching promising subspaces, certain search directions are identified *a priori* as those that cannot lead to an improved outcome. Examples include dynamic programming (DP) CSG algorithms [7] in which it is evaluated whether a decomposition a coalition C into exactly two smaller coalitions of all the agents in C would be profitable. In this spirit, the improved dynamic programming (IDP) algorithm presented in [8] considers more dissociations than just the dissociation of a coalition into two disjoint coalitions. This approach, which is yet another application of Lemma 1, turns out to be more efficient in terms of both time and memory costs than conventional DP algorithms.

Consider a system of agents where a value of a non-trivial coalition $C := \{a_1, a_2, a_3, a_4\}$ is 7. In order to prove that C cannot be in an optimal CS, it is not always necessary to show that any partition of this coalition has a combined value greater than 7. In the spirit of Lemma 1, it may also be sufficient to show that the values of a strict subset of k disjoint coalitions are greater than $v(C)$. For instance, it may be the case that $v(\{a_1\}) + v(\{a_3, a_4\}) \geq v(\{a_1, a_2, a_3, a_4\})$. Following this intuition, we can relax some of the assumptions in Lemma 1 as follows:

Lemma 2. *For any non-trivial coalition C that can be divided into k disjoint subcoalitions $C_1, \ldots, C_k$ where $C_1 \cup C_2 \ldots \cup C_k = C$; if $\sum_{i=1}^{j} v(C_i) > v(C)$ where $j < k$ then $C \notin CS^*$ and so $\forall CS : C \in CS, CS$ does not belong to $\{CS^*\}$.*

Theoretically, every feasible non-trivial coalition could be decomposed into all possible combinations of sub-coalitions; however, this is a completely inefficient approach. After all, such a decomposition of the grand coalition yields Π. In the remainder of this paper, we will show that an appropriate application of both Lemmas 1 and 2 may still considerably speed up the CSG process in the state-of-the-art CSG algorithm.

Firstly, we will extend Lemma 1 so that it can be applied not only to a particular coalition but to collections of coalitions which have been grouped together w.r.t. some criteria. One natural criterion to group coalitions is size. Let all coalitions of the same size be grouped together in $|A|$ sets $\mathcal{C}_1, \ldots, \mathcal{C}_i, \ldots, \mathcal{C}_{|A|}$ where $\mathcal{C}_i$ denotes the set of all coalitions of size i. For example, for $A = \{1, \ldots, 5\}$ there will be $|A| = 5$ sets $C_1, \ldots, C_5$ where C_1 contains all coalitions of size 1, C_2 all coalitions of size 2, *etc.* Additionally, in this example, suppose that the coalitions with smallest values in C_1 and C_2 are $\{a_1\}$ and $\{a_2, a_3\}$, respectively. This means that any decomposition of coalition of size 3 will not have a smaller value than $v(\{a_1\}) + v(\{a_2, a_3\})$. Consequently, if any coalition from C_3 has a value smaller than $v(\{a_1\}) + v(\{a_2, a_3\})$ then we can disregard this coalition as, following Lemma 1, it cannot be in an optimal structure. We extend Lemma 1 as follows:

Lemma 3. *Let C_i denote the (complete) set of all coalitions of size i. For any set $Z_i \subseteq C_i$ of coalitions of size i and for a particular integer partition $p = i_1, \ldots, i_k$ of i, (i.e., $i_1 + \ldots + i_k = i$) if the sum of the lowest coalition values in sets $C_{i_1}, \ldots, C_{i_k}$ (denoted $d_i(p)$) is strictly greater than the maximum value in set Z_i then no coalition from set Z_i can be in an optimal structure. More formally, if $d_i(p) := \sum_{i=1}^{k} \min C_{i_k} > \max Z_i$ then $\forall CS : C \in CS$ and $C \in Z_i$ it holds that $CS \notin \{CS^*\}$.*

Proof. Suppose that coalition C_i is the coalition with the biggest value in set $Z_i \subseteq C_i$ and coalitions $C_{i_1}, \ldots, C_{i_k}$ are the coalitions with the smallest value in lists $C_{i_1}, \ldots, C_{i_k}$, respectively. Now consider any coalition structure CS which contains coalition C_i. If CS is an optimal structure no other structure can have value greater than this. Now, consider structure $CS' = CS \setminus \{C_i\} \cup \{C_{i_1} \cup \ldots \cup C_{i_k}\}$ where $C_{i_1}, \ldots, C_{i_k}$ are all disjoint and $C_{i_1} \cup \ldots \cup C_{i_k} = C_i$. If the sum of the smallest values in sets $C_{i_1}, \ldots, C_{i_k}$ is greater than the biggest value in set $Z_i \subseteq C_i$ then clearly for all coalitions $C_{i_1}, \ldots, C_{i_k}$ in $C_{i_1}, \ldots, C_{i_k}$ respectively, and any $C_i \in Z_i \subseteq C_i$, $v(C_{i_1}) + \ldots + v(C_{i_k}) > v(C_i)$ and so $v(CS') > v(CS)$. Therefore, following Lemma 1, for any partition of i, $i_1, \ldots, i_k$ such that $i_1 + \ldots + i_k = i$, if the sum of the minimum values in sets $C_{i_1}, \ldots, C_{i_k}$ is greater than the maximum value in set $Z_i \subseteq C_i$ then no coalition in $Z_i \subseteq C_i$ can be in an optimal coalition structure and so no CS which contains a coalition in $Z_i \subseteq C_i$ can belong to $\{CS^*\}$.

For any set containing non-trivial coalitions of size i, it is possible to compute the set of all integer partitions of value i. Such a set is denoted $P(i)$. Furthermore, for each $p' \in P(i)$ it is possible to compute a value $d_i(p')$ as in Lemma 3. Now, following the same lemma, we can compare every coalition C of size i with $d_i(p')$ and immediately disregard those for which $v(C) < d_i(p')$. In fact, there is no need to apply Lemma 3 to all partitions in $P(i)$ but only to the partition $p'' \in P(i)$ such that $d_i(p'')$ is maximal in $P(i)$. Clearly, for a coalition C of size i, if it holds that $v(C) < d_i(p')$ then, $\forall p'' \in P(i) \setminus \{p'\}$ it also holds that $v(C) < d_i(p') \leq d_i(p'')$. Such a maximal value will be referred to as the *domination value* of coalitions of size i. More formally:

Definition 1. *For any set containing coalitions of size i and every partition $p \in P(i)$, the* domination value *$\tilde{d}_i$ is the highest value of $d_i(p)$ for all $p \in P(i)$, or $\tilde{d}_i = \max_{p \in P(i)} d_i(p)$.*

Following Lemma 1, if the value of any coalition C of size i is less than the domination value $\tilde{d}_i$ then there exists a dissociation of C with a value greater than $v(C)$. In such a case, C cannot be in any optimal structure. This property is exploited by the filter rules presented in the next section.

3   Filter Rules

In this section, we propose filter rules that can be applied both while calculating the values of all the coalitions in F (thus, generating $V(F)$) and while searching through the (sub-)spaces of coalition structures. We will refer to coalitions which cannot belong to any optimal structure as *not-promising*. All the other coalitions will be called *promising*. We denote both of these disjoint sets by $F_{np} \subseteq F$ and $F_p \subseteq F$, respectively. Initially, before the filter rules are applied, it is assumed that all coalitions are promising, *i.e.*, $F_p = F$. Furthermore, we will refer to a subset of coalition values $Z \subseteq V(F_p)$ ($Z \subseteq V(F_{np})$) as promising (not promising).

3.1   Filter Rules for Input Calculation

While calculating all coalition values in the input, Lemma 2 can be applied to highlight those coalitions which are not promising. However, dissociating every coalition into all potential sub-coalition combinations is usually inefficient for coalitions of greater size. It should be left to a system designer to decide into how many sub-coalitions a currently computed coalition should be disaggregated. The natural and possibly most efficient choice is to consider the singleton partition of a coalition, *i.e.*, the decomposition of the coalition into all of the individual agents who participate in this coalition. Clearly, in the process of computing the value of a given coalition, all agents who co-operate in this coalition must be known and such a disaggregation can be easily performed. Therefore, following Lemma 2 we propose the first filter rule:

FR1. For any non-trivial coalition C that can be divided into k disjoint singleton sub-coalitions $C_1, \ldots, C_k$ where $C_1 \cup C_2 \ldots \cup C_k = C$ and $\forall i = 1, \ldots, k$ $|C_i| = 1$; if it is the case that either (i) $v(C_1) + \ldots + v(C_k) > v(C)$ or (ii) $\sum_{i=1}^{j} v(C_i) > v(C)$ where $j < k$ then all such coalitions $C \in F_{np}$.

Naturally, we can relax the constraint that $|C_i| = 1$ to $|C_i| \leq s$ where $1 < s < |A|$ depending on into how many partitions the system designer wishes to divide coalition C. The computational cost of applying **FR1** should be balanced against potential gains.

Example 1. Consider a five agent system $A = \{a_1, a_2, a_3, a_4, a_5\}$ with the following coalition values: $v(C) = 14$ for $C = A$; $\forall |C| = 1$ $v(C) = 3$; $\forall |C| = 2$ $v(C) = 5$; $\forall |C| = 3$ $v(C) = 10$;, and $\forall |C| = 4$ $v(C) = 12$. Observe that the sum of any two coalitions of size 1 is strictly greater than the value of any coalition of size 2. Furthermore, the value of the grand coalition is smaller than the value of a non-cooperative coalition structure. Thus, **FR1** filters out all coalitions of size two as well as the grand coalition. However, following Lemma 1 the system designer may consider all dissociations of a coalition into exactly two disjoint sub-coalitions in the same way as in the DP algorithm presented in [7]. Since the value of any coalition of size 4 is smaller than the value of any coalition of size 1 added to the value of a disjoint coalition of size 3, then coalitions of size 4 are not promising.

Now, consider the domination value. Lemma 3 immediately yields the following filter rule:

FR2. For any subset $Z_s \subseteq C_s$ of coalitions of size s, if $\tilde{d}_s > \max Z_s$ then all coalition from Z_s are not promising. More formally: $\forall Z_s \subseteq C_s$, if $\tilde{d}_s > \max Z_s$ then $Z_s \in F_{np}$ and $Z_s \notin F_p$.

FR2 can be applied by a system designer to every coalition C of size i immediately after $v(C)$ has been calculated (*i.e.* in the same manner as **FR1**). Alternatively, if $|Z_i| \geq 2$ then the value of $\max Z_i$ can be recorded while calculating coalition values in this subset and **FR2** can be applied after this process has been finished. We illustrate the functioning of **FR2** with the following example:

Example 2. Suppose the coalition values for four agents $A = \{a_1, a_2, a_3, a_4\}$ are as follows: $\forall |C| = 1, v(C) \in \langle 4, 7 \rangle, \forall |C| = 2, v(C) \in \langle 5, 7 \rangle, \forall |C| = 3, v(C) \in \langle 7, 11.5 \rangle$ and $v(A) = 11$.[3] The domination values for the coalitions of size 2 to 4 are computed as follows: $\tilde{d}_2 = \min S_1 + \min S_1 = 4 + 4 = 8, \tilde{d}_3, = \max\{3 \times \min S_1, \min S_1 + \min S_2\} = 12, \tilde{d}_4, = \max\{4 \times \min S_1, 2 \times \min S_1 + \min S_3, 2 \times \min S_2\} = 16$. Observe that both $\tilde{d}_2 > \max S_2$ and $\tilde{d}_4 > \max S_2 = v(A)$. Intuitively, this means that for every coalition of size 2 and 4, there exists a dissociation of this coalition into sub-coalitions which have value greater than the value of the coalition and so following previous reasoning, no coalition of size 2 or 4 can be in an optimal structure and no structure containing these coalitions can be in an optimal set.

3.2   Filter Rules for Search of Coalition Structure Space

As mentioned in the Introduction, the anytime algorithm of Rahwan *et al.* divides Π into sub-spaces containing structures which are identical w.r.t. size of the coalitions involved. In a system of $|A|$ agents, let $S^* := \{m_1, \ldots, m_k\}$ denote the (currently) most promising subspace, where $m_1, \ldots, m_k$ represent sizes of the coalition involved ($\sum_{i=1}^{k} m_i = |A|$ and $k \geq 2$). To search S^*, the values of all the coalition structures belonging to this sub-space should be computed unless it is proven beforehand that they cannot belong to $\{CS^*\}$.[4] The general form of a coalition structure in S^* is $\{C_1, \ldots, C_k\}$ such that all of $C_1, \ldots, C_k$ are disjoint and $\forall i = 1, \ldots, k$ $C_i \in C_{m_i}$. Let CS_N^* denote the coalition structure with the highest value found thus far. Rahwan *et al.* propose a filter rule that, based on the statistical information gathered about the coalition value input, avoids those structures which cannot be optimal. For $k \geq 3$ and $k - 1 \geq l \geq 1$ it holds that:

B&B. If $\sum_{i=1}^{l} v(C_i) + \sum_{i=l+1}^{k} \max C_{m_i} \leq v(CS_N^*)$ then no structures in S^* can be optimal, to which simultaneously belong all of $C_1, \ldots, C_l$.

This filter rule ensures that, for a particular structure under consideration $\{C_1, \ldots, C_k\}$, if the combined value of first $1 \leq l \leq n - 1$ coalitions ($\sum_{i=1}^{l} v(C_i)$) plus the value of the sum of maximum values of coalitions in the remaining sets $C_{m_{l+1}}, \ldots, C_{m_k}$ is less than the current optimum value $v(CS_N^*)$ then no structures to which all of $C_1, \ldots, C_l$ simultaneously belong will be considered in the optimal CSG process.

---

[3] For example, the notation $v(C) \in \langle 4, 7 \rangle$ means that $v(C)$ can be any real value higher than or equal to 4 and lower than or equal to 7 and that the minimal value of all such coalitions is 4 and the maximal is 7.

[4] Or unless it is proven that an optimal coalition structure in this sub-space has been found.

Example 3. For $|A| = 9$ agents let $S^* := \{1, 2, 2, 2, 2\}$ be the (currently) most promising subspace, $CS_N^* = 28$, $v(\{a_1\}) + v(\{a_2, a_3\}) = 9$, $v(\{a_4, a_5\}) = 4$, $v(\{a_4, a_6\}) = 6$ and $\max \mathcal{C}_2 = 7$. Following **B&B**, since $v(\{a_1\}) + v(\{a_2, a_3\}) + v(\{a_4, a_5\}) + \max \mathcal{C}_2 + \max \mathcal{C}_2 < CS_N^*$ then no structures in S^* can be optimal, to which simultaneously belong all of $\{1\}, \{2, 3\}$ and $\{4, 5\}$. These structures are: $\{\{a_1\}, \{a_2, a_3\}, \{a_4, a_5\}, \{a_6, a_7\}, \{a_8, a_9\}\}$, $\{\{a_1\}, \{a_2, a_3\}, \{a_4, a_5\}, \{a_6, a_8\}, \{a_7, a_9\}\}$, and $\{\{a_1\}, \{a_2, a_3\}, \{a_4, a_5\}, \{a_6, a_9\}, \{a_7, a_8\}\}$. Thus, in this example branch-and-bound rule saves on calculation time by avoiding calculations that lead to three structures which cannot be optimal.

The above branch-and-bound technique is based on basic statistical information collected about $\mathcal{C}_m$, namely the maximum value of coalitions in this set ($\max \mathcal{C}_m$). Assuming that this information is known about some subsets $Z_m \subseteq \mathcal{C}_m$ for $m = m_1, \ldots, m_k$ then the filter rule above can be generalized as follows:

FR3. If $\sum_{i=1}^{k} \max Z_{m_i} < v(CS_N^*)$, where $k \geq 2$, then no structures in S^* can be optimal, in which simultaneously $C_1 \in Z_{m_1}, \ldots$ and $C_k \in Z_{m_k}$.[5]

Example 4. In the system of nine agents let $S^* := \{1, 2, 2, 4\}$ be the (currently) most promising subspace, $CS_N^* = 25$, $\max\{Z_1 := \{\{a_1\}, \{a_2\}, \{a_3\}\}\} = 4$, $\max\{Z_2 := \{\{a_2, a_3\}, \{a_4, a_5\}, \{a_6, a_7\}, \{a_8, a_9\}, \{a_1, a_3\}, \{a_1, a_2\}\}\} = 6$ and $\max\{Z_4 := \mathcal{C}_4\} = 7$. Following **FR3**, since $\max Z_1 + 2 \times \max Z_2 + \max Z_4 < CS_N^*$ then no structures in S^* containing all of C_1, C_2, C_3, C_4 such that $C_1 \in Z_1$, $C_2 \in Z_2$, $C_3 \in Z_2$, and $C_4 \in Z_4$ can be optimal. In this example **FR3** saves on calculation time by avoiding $\{\{a_1\}, \{a_2, a_3\}, \{a_4, a_5\}, \{a_6, a_7, a_8, a_9\}\}$, $\{\{a_2\}, \{a_1, a_3\}, \{a_4, a_5\}, \{a_6, a_7, a_8, a_9\}\}$, and $\{\{a_1, a_2\}, \{a_3\}, \{a_4, a_5\}, \{a_6, a_7, a_8, a_9\}\}$. Note that there are certain combinations of $C_1 \in Z_1, C_2 \in Z_2, C_3 \in Z_2, C_4 \in Z_4$ for which $\{C_1, C_2, C_3, C_4\}$ is not a coalition structure. For instance, any combination $\{\{a_1\}, \{a_2, a_3\}, \{a_4, a_5\}, \{a_1, a_2, a_3, a_4\}\}$ is neither exhaustive nor disjoint. Although, we are not interested in such combinations, it can be observed that the inequality in **FR3** holds for them as well.

4   Application

To test the effectiveness of the filter rules we employ them in the state-of-the-art distributed coalition value calculation algorithm of Rahwan and Jennings [1] (DCVC from now on) and in the state-of-the-art anytime CSG algorithm of Rahwan *et al.* [2] (RCSG from now on).

Apart from the decentralised design, the key advantage of DCVC is that calculated coalition values in $V(F)$ are ordered in a unique way ensuring that only $V(F)$ and not F must be kept in memory. Coalition values structured in the same way as in DCVC are used as an input to RCSG. However, it is not trivial to connect both algorithms as the former one is distributed whereas the latter one is centralised. This means that, at the

---

[5] Note that certain combination of $C_1 \in Z_{m_1}, \ldots, C_k \in Z_{m_k}$ are neither disjoint nor exhaustive, *i.e.*, they are not proper coalition structures as assumed in this paper. We leave them aside as not being relevant to our analysis. See Example 4.

moment when calculations in DCVC are complete, every agent knows only a fraction of $V(F)$ needed in RCSG. Thus, a resource-consuming data transfer has to take place.

An application of any filter rules always requires a balance between computational costs and potential gains. We will show that filter rules **FR1**, **FR2** and **FR3** can be applied at a relatively low computational cost, and that using them results in:

1. A substantial decrease in the number of coalitions to be transferred; and
2. An increase in the efficiency of the promising structure sub-space search in RCSG.

4.1   Application of FR1, FR2 and FR3 in DCVC

In DCVC, the space of all coalitions is represented as a set of lists $L_1, \ldots, L_{|A|}$ where list L_i contains all coalitions of size i. Within each list, the coalitions are ordered w.r.t. the agents they compromise. This ordering ensures that the agent composition of every coalition in each list can be derived from the place in the list it occupies. In terms of the notation introduced in Section 2 we may write $L_i = \overrightarrow{C_i}$, i.e., every list is an ordered set of the values in C_i. Such lists for a system of six agents $a_1, a_2, ..., a_6$, all with equal computational capabilities, are presented in Figure 1.[6]

All lists are divided into disjoint segments proportional to the agents' computational capabilities. Usually every agent is assigned two segments, one in the upper and one in the lower part of the list. We will denote both segments assigned to agent a_i in list L_m by $\mathcal{L}_m^U(a_i)$ and $\mathcal{L}_m^L(a_i)$, respectively. As not all the lists are exactly divisible by $|A|$, variable α, depicted in Figure 1, is used to distribute 'left over' coalitions as equally as possible between agents. 'Left over' coalition values assigned to agent a_i will be denoted by $\mathcal{L}_m^O(a_i)$. Overall, the above methods to distribute $V(F)$ have the advantage that all agent computations are finished (almost) simultaneously, even when some coalitions require more arithmetic operations than others and when agents have different processing capabilities. The allocation process is described in detail in [1].

After the allocation of segments has been performed, agents sequentially compute values in lists starting from L_1. Since there are $|A|$ coalitions in L_1, each agents is assigned exactly one value from this list to calculate. Let us assume that every agent transmits this value to all the other agents in the system. By doing so, agents are ready to apply **FR1** while computing coalition values in L_2. Additionally, they are able to efficiently compute the value of a structure consisting of every promising coalition C and singletons, i.e., $\{C, \{j\} : j \in A \backslash C\}$.

For example, in Figure 1, while computing the values of coalitions $\{a_4, a_5\}$ and $\{a_3, a_6\}$■ in $\mathcal{L}_2^U(a_5)$, agent a_5 applies **FR1** and compares $v(\{a_4\}) + v(\{a_5\})$ with $v(\{a_4, a_5\})$ as well as $v(\{a_3\}) + v(\{a_6\})$ with $v(\{a_3, a_6\})$. Furthermore, this agent calculates the values of $\{\{a_1\}, \{a_2\}, \{a_3\}, \{a_4, a_5\}, \{a_6\}\}$ and $\{\{a_1\}, \{a_2\}, \{a_3, a_6\}, \{a_4\}, \{a_5\}\}$. Let $CS_N^*(a_i)$ denote a coalition structure with the highest value that agent a_i found in its assigned segments.

---

[6] Note that in Figure 1 numbers in lists are a shorthand notation representing agents. It should be emphasized that every list contains coalition values from $V(F)$ and not coalitions from F themselves. Thus, the representation of lists $L_1, ..., L_{|A|}$ in Figure 1 should be read not as coalitions of agents from F but as their numerical values from $V(F)$.

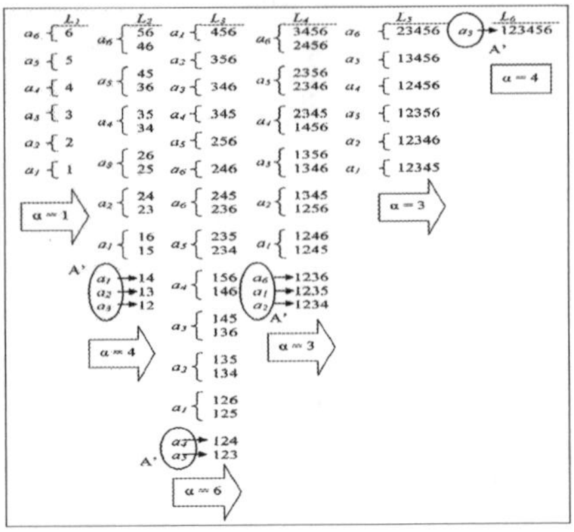

Fig. 1. Division of $V(F)$ in a system of six agents

In DCVC, when the computational capabilities of all the agents are symmetric, each of them receives an (almost) equal fraction of F to calculate $V(F)$ (as in Figure 1). The total number of values assigned to every agent is equal to either $\lfloor (2^{|A|} - 1)/|A| \rfloor$ or $\lfloor (2^{|A|} - 1)/|A| \rfloor + 1$, depending on the distribution of 'left-over' coalitions. For example, for 20 agents this amounts to either 52,428 or 52,429 and for 30 agents to either 53,687,091 or 53,687,092. Clearly, transferring such numbers of coalition values from DCVC to RCSG is a potential bottleneck in the system performance. The application of **FR1** helps to reduce the transfer load but its effectiveness depends on the values of the singletons. In some systems, these values might be lower than a value of any coalition. In such a case, following Lemma 2, **FR1** can be extended to consider other partitions of a coalition C whose value is being computed. However, dissociations other than into singletons may considerably increase computational cost and should be weighted against potential gains. Consequently, we propose another approach that can significantly reduce the transfer load. Since RCSG searches only through promising subspaces of Π, then only some lists of coalition values have to be transferred. This means that the procedure to evaluate whether a given subspace is promising should be removed from the RCSG and incorporated into DCVC.

In RCSG a subspace S is considered to be promising if its upper bound $UB_S := \sum_{\forall m_i \in S} maxL_{m_i}$ is no smaller than the lower bound of the entire system $LB := \max\{v(CS_N^*), \max\{Avg_S\}\}$, where $Avg_S = \sum_{\forall m_i \in S} avgL_{m_i}$ is proven to be the average value of every subspace. In DCVC, where calculations are distributed among all of the agents, statistics such as maximum, minimum and average coalition values cannot be computed without an exchange of information among agents. However, if agents record maximum, minimum and average values in every segment they are assigned to calculate then after all calculations are (almost simultaneously) finished every agent a_i can broadcast this statistical data along with $CS_N^*(a_i)$. Using this knowledge, the agents can calculate, for every list L_m, the values of $maxL_m$, $avgL_m$, and $minL_m$. Both the maximum and average values are then used, by the agents, to determine which sub-spaces are promising. Consequently, by exchanging some basic information about segments, the agents can prune the search space at the end of DCVC in the same way as the centre would do it in RCSG.

The subspace with the maximum upper bound is the most promising, *i.e.*, $S^* = \arg\max_S UB_S$, and is searched first. It might happen that the unique CS^* (*i.e.*, when $|\{CS^*\}| = 1$) is found in this sub-space. Thus, in order to reduce potential transfer load, we assume that only lists needed to construct coalition structures in S^* are initially passed on from DCVC to RCSG. Let the most promising subspace be $S^* := \{m_1, \ldots, m_k\}$. While searching all structures in this subspace, a natural assumption is that each agent will begin their search, starting from the segment of values allocated to it in L_{m_1}. For instance, let $S^* = \{1, 2, 3\}$ for $n = 6$. Referring to Figure 1, it would be natural to assume that agent a_1 generates all the coalition structures starting with coalition $\{a_6\}$, agent a_2 generates those starting with $\{a_5\}$, *etc.* However, this means that the coalitions in the two longer lists L_2 and L_3 must be transferred between agents. Therefore, one way to minimize the transfer load is to assume agents start searching S^* beginning with the longest list. In our example, let S^* be defined not as $\{1, 2, 3\}$ but as $S^* = \{3, 1, 2\}$. With this ordering, agent a_1 generates all the coalition structures starting with coalitions $\{a_4, a_5, a_6\}$, $\{a_1, a_2, a_6\}$ and $\{a_1, a_2, a_5\}$. Thus, values from L_3 need not be transferred as agents utilize only the values assigned to them in the DCVC. Therefore, in contrast to [9], such an ordering, which starts with the longest list in S^*, will be the one considered in our algorithm. This means agents do not have to transfer values in L_{m_1}. However, if L_{m_1}, which is the longest list in S^*, appears more than once then the list must be transferred as it will be needed by all the agents during the search process.[7]

To reduce the transfer load even further, both **FR2** and a version of **FR3** are applied. Before the transmission begins, agents calculate the domination value $\tilde{d}_m$ for every list $L_m : m \in S^*$ using the relevant statistics.[8] Segments assigned to agent a_i which meet the condition $\max \mathcal{L}_m^{U/L/O}(a_i) < \tilde{d}_m$ are filtered out as not promising. Furthermore, for all segments in which $\max \mathcal{L}_m^{U/L/O}(a_i) \geq \tilde{d}_m$, it is determined if individual coalition values within these segments, not filtered by **FR1** are not smaller than $\tilde{d}_m$. Individual coalitions with such values become not promising. The version of **FR2** that concerns segments will be denoted as **FR2a**, whereas, the version that concerns individual coalitions as **FR2b**.

Although **FR3** has been constructed to increase efficiency of searching Π, a version of this filter rule can also be applied before data transfer. In S^*, every agent a_1 knows $\max \mathcal{L}_m^{U/L/O}(a_i)$ and $\max L_m$ for all $m = m_1, \ldots, m_k$ the following version of **FR3** (denoted as **FR3a**) can be applied to decide whether either $\mathcal{L}_{m_j}^U(a_i)$, $\mathcal{L}_{m_j}^L(a_i)$ or $\mathcal{L}_{m_j}^O(a_i)$, where $m_j \in S^*$, can be filtered out as not promising:

$$\sum_{m \in S^* \setminus \{m_j\}} \max L_m + \max \mathcal{L}_{m_j}^{U/L/O}(a_i) < v(CS_N^*). \tag{1}$$

---

[7] For more details on the search process, see Subsection 4.2.

[8] We leave it to the discretion of the system designer as to the exact architecture of domination value calculations. At the end of DCVC, agents can either calculate all needed domination values individually or distribute calculations among themselves in such a way that the most efficient agents calculate domination values for the lists of coalitions with the highest cardinality.

If the segment cannot be filtered out as not promising then the above filter rule can be applied to every individual coalition C within this segment not filtered out by **FR1** or **FR2a/b**. In such a situation, $\max \mathcal{L}_{m_j}^{U/L/O}(a_i)$ in formula (1) can be replaced with $v(C)$, where $v(C) \in \mathcal{L}_{m_j}^{U/L/O}(a_i)$ and this version of **FR3** is denoted **FR3b**. Finally, after all the filter rules have been applied, the agents transfer only those values which are promising.

To summarize, in order to reduce the transfer load from DCVC to RCSG we propose the following extension of the former algorithm:

Step 1: Agents exchange among themselves values of singleton coalitions in L_1. While calculating value of any non-trivial coalition C they: (1.1) apply **FR1**; (1.2) compute $CS_N^*(a_i)$ and (1.3) Update and store the statistics about their segments $\mathcal{L}_m^{U/L/O}$;

Step 2: After calculating the values in all of their segments, agents exchange among themselves the segments' statistics and $CS_N^*(a_i)$. Using this information, they determine the most promising sub-space S^* and domination values $\tilde{d}_m$ for every list $L_m : m \in S^*$ that needs to be transmitted;

Step 3: Just before transmission takes place each agent applies both **FR2a** and **FR3a** to segments $\mathcal{L}_m^{U/L/O}$, where $m \in S^*$. If a segment is not filtered out then **FR2b** and **FR3b** are applied again to individual coalitions within this segment that have not been filtered out by **FR1**. Only the values of promising coalitions are transferred.[9] If the longest list in S^* appears only once then this list is not transferred.

Finally, it should be observed that even if the coalition value calculation algorithm was not distributed, then **FR1**, **FR2b** and **FR3b** can be still be applied to determine the not promising coalitions. In the simulations we will show that it considerably reduces the running time of RCSG.

4.2 Application of FR3 in RCSG

The strength of RCSG in searching the promising sub-spaces is that it avoids creating both invalid and repeated coalition structures as described in [9]. Suppose that $S^* := \{m_1, \ldots, m_k\}$ is the most promising subspace in a system of $|A|$ agents and that the value of the particular partial structure $\{C_{m_1}, \ldots, C_{m_l}\}$, where $l < k$, has already been calculated. Let $\overrightarrow{A'}$ be the ordered set of agents which do not belong to any coalition in this partial structure, i.e. $\overrightarrow{A'} := \overrightarrow{A} \backslash \{C_{m_1}, \ldots, C_{m_l}\}$. The method proposed by Rahwan et al. in [9] cycles through all m_{l+1}−element combination from $\overrightarrow{A'}$ to construct all feasible coalitions $C_{m_{l+1}}$. The cycle is designed in such a way that all m_{l+1}−combinations which must contain the agent in $\overrightarrow{A'}$ with the lowest index are considered first and all m_{l+1}−combinations which must contain the second agent in $\overrightarrow{A'}$ but not the first one are considered in the next step and so on. As explained in

---

[9] To indicate the position of each promising coalition value and maintain the list structure we propose to transmit a characteristic bit vector alongside the stream of values. In such a vector 1 indicates value of a promising coalition and 0 a not promising one.

Subsection 3.2, after constructing every $C_{m_{l+1}}$ RCSG applies **B&B** filter rule to decide whether any partial structure $\{C_{m_1}, ..., C_{m_l}, C_{m_{l+1}}\}$ is promising. A version of **FR3** can be used to identify groups rather than only individual partial structures. To this end, agents in DCVC have to gather additional information about some particular segments in lists $L_1, ..., L_{|A|}$. Let $\overrightarrow{Z}_m(i)$ denote a segment of list L_m such that a_i is the first agent in every coalition in $\overrightarrow{Z}_m(i) \subseteq L_m$. For example, referring to Figure 1, $\overrightarrow{Z}_3(2)$ is $\{\{2,5,6\}, \{2,4,6\}, \{2,4,5\}, \{2,3,6\}, \{2,3,5\}, \{2,3,4\}\}$. Assume that, while calculating coalition values in DCVC, the agents record $\max \overrightarrow{Z}_m(i)$ for every combination of $m = 1, ..., |A|$ and $a_i \in A$. In such a case, before cycling through all m_{l+1}−combinations which contain the agent in $\overrightarrow{A'}$ with the lowest index, the following version of **FR3** (denoted as **FR3c**) can be applied, by a centre in RCSG. If

$$\sum_{i=1}^{l} v(C_i) + \max \overrightarrow{Z}_{l+1}(i) + \sum_{i=l+2}^{k} \max L_{m_i} < v(CS_N^*),$$

then any partial structure $\{C_{m_1}, ..., C_{m_l}, C_{m_{l+1}}\}$ such that $C_{m_{l+1}} \in \overrightarrow{Z}_{l+1}$ is not promising. **FR3c** is a generalization of RCSG **B&B** as it is applicable to groups rather than individual coalitions.

4.3 Numerical Simulations

In the numerical simulations we compare the sequential execution of DCVC and RCSG with and without filter rules. The following assumptions are imposed: (i) the calculation of any particular coalition values in DCVC takes no time; and (ii) any data transfers are instantaneous. We focus on the percentage of $V(F)$ which does not have to be transmitted from DCVC to RCSG due to applied filter rules. Additionally, we demonstrate that, for a normal distribution of coalition values, filter rules considerably improve the performance of the CSG process in RCSG.

Following [2,5], we evaluate the algorithms under two different assumptions as to the probability distributions of coalition values:[10] (ND) Normal: $v(\mathbf{C}) = max(0, |C| \times p)$, where $p \in N(\mu = 1, \sigma = 0.1)$; and (UD) Uniform: $v(\mathbf{C}) = max(0, |C| \times p)$, where $p \in U(a, b)$, where $a = 0$ and $b = 1$. For a system of $|A| = 11 ... 20$ agents, we ran both versions of the algorithms 25 times and reported the results within a 95% confidence interval. The algorithms were implemented in MATLAB.

The results are presented in Table 1. Column 2 shows the number of all coalition values $|F|$ and Column 3 percentage of $V(F)$ that needs to be transmitted from DCVC to RCSG without filter rules. Column 4 shows the actual percentage of $V(F)$ transferred when filter rules are applied. Column 5 contains the size of the characteristic bit vector (as a percentage of $V(F)$) that must be transferred simultaneously with the values in Column 4 (see Footnote 8). The next five columns present the individual contribution of filter rules **FR1**, **FR2a**, **FR2b**, **FR3a** and **FR3b** to the reduction of the overall transfer load between Columns 3 and 4. The last column shows the running time of RCSG based on the unfiltered input and without **FR3c** divided by the running time of this algorithm based on filtered input and with **FR3c**.

---

[10] We omit the sub- and super-additive cases as their solution is straightforward.

Table 1. Simulation results (all values, except for columns 1,2, and 11 are expressed in %)

1.	2.	3.	4.	5.	6.	7.	8.	9.	10.	11.
	A		V(F)	without FRs	with FRs	c. hit vector	FR1	FR2a	FR2b	FR3a	FR3b	Relative Time
11	2047	91±4.5	8.68±1.69	5.61		53.12	0	0.01	21.61	25.26	2.75±0.83
12	4095	91±2.4	8.43±1.51	5.55		55.71	0	0.01	18.13	26.15	4.17±1.55
13	8191	90±3.5	5.29±1.37	5.5		54.91	0	0	18.69	28.4	6.61±2.03
14	16383	91±3.3	6.05±1.61	5.7		54.99	0	0	16.7	28.31	16.31±5.0
15	32767	93±8.4	5.04±1.21	5.9		53.11	0	0	14.7	32.16	17.88±6.04
16	65535	91±5.1	4.48±1.14	4.48±1.14		49.79	0	0	14.04	36.17	23.21±7.4
17	131071	92±4.5	4.29±1.01	4.29±1.01		53	0	0	11.19	35.81	112.2±4.4
18	262143	93±6.3	3.69±0.89	3.69±0.89		52.47	0	0	9.01	38.52	169±54
19	524287	94±6.1	3.12±0.81	3.12±0.81		51.88	0	0	7.8	40.72	197±67
20	1048575	91±5.3	2.49±0.67	2.49±0.67		50.05	0	0	5.99	43.95	380±88
	A		V(F)	without FRs	with FRs	c. hit vector	FR1	FR2a	FR2b	FR3a	FR3b	Relative Time
11	2047	51.31±6.51	3.32±1.79	3.32±1.79		49.24	0.07	1.17	35.21	14.31	1.12±0.14
12	4095	39.99±18.61	2.89±1.41	2.89±1.41		51.3	0.06	1	28.01	19.63	1.22±0.09
13	8191	49.68±17.52	2.71±1.02	2.71±1.02		49.43	0.12	0.68	27.83	21.94	1.34±0.18
14	16383	41.13±18.80	2.42±0.68	2.42±0.68		50.32	0.04	0.32	23.15	26.17	1.14±0.12
15	32767	48.52±21.60	2.35±0.48	2.35±0.48		51.12	0	0.11	18.51	30.26	1.23±0.16
16	65535	52.98±19.40	1.26±0.33	1.26±0.33		50.51	0	0.04	16.54	32.91	1.14±0.09
17	131071	50.41±18.70	1.15±0.29	1.15±0.29		51.31	0	0.02	11.19	36.98	1.01±0.03
18	262143	46.41±19.70	1.13±0.28	1.13±0.28		50.18	0	0	10.91	38.91	1.03±0.02
19	524287	41.71±18.42	1.05±0.13	1.05±0.13		50.02	0	0	8.01	41.97	1.04±0.04
20	1048575	32.80±19.70	1.04±0.1	1.04±0.1		50.25	0	0	7.46	43.29	1.19±0.11

Consider ND first. Without filter rules, about 90% of $V(F)$ needs to be transmitted from DCVC to RCSG (Column 3). Filter rules reduce this number to about $9\% + 5.5\%$ for $n = 11$ and to around $2.5\% + 2.5\%$ for $n = 20$ (Columns 4 and 5).[11] **FR1** is the most successful in filtering coalitions accounting for about 50% of them. Consequently, both **FR2a** and **FR2b**, based on the same assumption, cannot offer any significant improvement upon this result. However, if the singleton coalitions have relatively low values, then **FR1** would not perform so well, and **FR2a/b** would become more profitable. A decreasing percentage of coalition values are ruled out by rule **FR3a**. The intuition behind the decreasing effectiveness of this filter rule is as follows. The higher the value of $|A|$, the longer the segments become. Consequently, there is a greater probability that the (randomly-drawn) extreme values in these segments are similar to each other. This reduces the effectiveness of filter rules based on segments' maxima. In contrast, **FR3b**, which focuses on the particular coalition values, has increasing success. We expect that for $n > 25$ only **FR1** and **FR3b** should be used. In conclusion, the combination of the filtered input and the application of **FR3c** in RCSG results in a much faster performance of this algorithm. Experiments show that this result is to be partially attributed to filter rules and primarily to the re-ordering of the way in which the subspaces are searched.[12] Intuitively, in the ND case, more coalitions are filtered from the longest lists than from any other. This means that the number of coalition structure to evaluate is already significantly decreased.

For the UD case, only about 50% of $V(F)$ needs to be transmitted from DCVC to RCSG (Column 3) but filter rules are able to reduce this number: from about $2 \times (3.32\% \pm 1.79\%)$ for $n = 11$ to around $2 \times (1.04\% + 0.1\%)$ for $n = 20$. Analysis of the effectiveness of individual filter rules is similar to the ND case. However, in contrast to ND, the combination of the filtered input and the application of **FR3c** does not significantly outperform the standard RCSG. This is caused by the very nature of the uniform-distribution. Intuitively, since coalition values in any lists are dispersed

---

[11] All simulations are performed in MATLAB. In the future, the same simulations will be reprogrammed in JAVA but this should not influence the results presented in this paper which are based on relative comparison.

[12] See Subsection 4.1 for more details.

neither reduced input nor **FR3c** perform much better than **B&B** in standard RCSG. However, our filter rules still prove its effectiveness by achieving large reduction in the transfer load.

5 Conclusions

In this paper we have discussed a number of techniques designed to increase the efficiency of CSG algorithms. In particular, we developed filter rules which can, for CFG representations, eliminate a significant proportion of not promising coalitions from the space of all coalition values. Such a (structured) space of coalition values acts as input to, among others, the state-of-the-art CSG algorithm of Rahwan *et al.* Although this algorithm has been demonstrated to be very effective, its particular drawback is its centralised nature. This is especially challenging as there exist already a very efficient algorithm for distributed coalition value calculation; this means that after the DCVC algorithm has been employed, all of the agents have to transmit all the values they have computed to a single entity. In this paper, we demonstrated that our proposed filter rules are extremely effective in reducing the size of this coalition value input to the CSG algorithm. Consequently, we showed how to efficiently bridge the gap between the decentralised DCVC and the centralised RCSG. A natural follow up to our work, which we plan to explore, is to develop a distributed CSG algorithm that can be relatively easily constructed from the analysis in this paper.

References

1. Rahwan, T., Jennings, N.: An algorithm for distributing coalitional value calculations among cooperating agents. Artificial Inteligence 171(8-9), 535–567 (2007)
2. Rahwan, T., Ramchurn, S., Dang, V., Giovannucci, A., Jennings, N.: Anytime optimal coalition structure generation. In: Proceedings of AAAI, Vancouver, Canada, pp. 1184–1190 (2007)
3. Michalak, T., Dowell, A., McBurney, P., Wooldridge, M.: Optimal coalition structure generation in partition function games. In: ECAI, Patras, Greece (2008)
4. Shehory, O., Kraus, S.: Methods for task allocation via agent coalition formation. Artificial Intelligence 101(1), 165–200 (1998)
5. Sandholm, T., Larson, K., Andersson, M., Shehory, O., Tohme, F.: Coalition structure generation with worst case guarantees. A.I. 111(1-2), 209–238 (1999)
6. Dang, V., Jennings, N.: Generating coalition structures with finite bound from the optimal guarantees. In: Proceedings of AAMAS, New York, USA (2004)
7. Yeh, D.Y.: A dynamic programming approach to the complete partition problem. BIT 26(4), 467–474 (1986)
8. Rahwan, T., Jennings, N.: An improved dynamic programming algorithm for coalition structure generation. In: Proceedings of AAMAS, Estoril, Portugal, (2008)
9. Rahwan, T., Ramchurn, S., Dang, V., Giovannucci, A., Jennings, N.: Near-optimal anytime coalition structure generation. In: Proceedings of IJCAI, Hyderabad, India (2007)

Cognitive Use of Artifacts: Exploiting Relevant Information Residing in MAS Environments

Michele Piunti[1,2] and Alessandro Ricci[2]

[1] Institute of Cognitive Sciences and Technologies, ISTC-CNR, Rome, Italy
[2] Alma Mater Studiorum, Università degli studi di Bologna, DEIS, Cesena, Italy
{michele.piunti,a.ricci}@unibo.it

Abstract. Besides using language and direct communication, humans adopt various kind of *artifacts* as effective means to represent, store and share information, and finally support knowledge-based cooperation in complex work environments. Similarly to the human case, we argue that an analogous concept can be effective also in the context of multi-agent systems (MAS) when cognitive agents are of concern. In particular, we investigate the use of *cognitive artifacts*, as those computational entities designed to store, process and make available environmental resources which are relevant to achieve goals and coordinate their cooperative and distributed activities. Some of the practical benefits of the artifact-based approach are discussed through an experiment based on CArtAgO and *Jason* technologies. Effectiveness of different interaction strategies are investigated for teams of goal-oriented agents using different kind of comunication styles.

1 Introduction

According to conceptual frameworks such as Activity Theory [6] and Distributed/Situated Cognition, *artifacts* play a fundamental role in human cooperative activities, deeply affecting individuals problem solving and ability to perform cooperative tasks, manage and share knowledge, cooperate and coordinate their work. In this view, artifacts are a way to conceive, structure and organise the *environment* where humans live and work. Recent agent literature highlighted the important role that the notion of environment can play in designing and engineering Multi-Agent Systems (MAS). The environment is thus conceived as the suitable locus to encapsulate resources, services and functionalities in order to improve agent interaction, coordination, and cooperation [13]. Not surprisingly, *coordination artifacts* – like calendars, programmable blackboards, schedulers, etcetera – have been progressively introduced in MAS for instance to enable and mediate agent-agent interaction and communication [9]. In this view, the notion of artifact has been recently introduced by the Agents & Artifact (A&A) conceptual model as first-class entity to structure and organise agent work environments, representing resources and tools[1] that agents may want to use to support their activities, analogously to artifacts in the human case [8].

---

[1] In this context we consider the terms *artifacts* and *tools* as synonyms.

J.-J.Ch. Meyer and J.M. Broersen (Eds.): KRAMAS 2008, LNAI 5605, pp. 114–129, 2009.
© Springer-Verlag Berlin Heidelberg 2009

Besides communication models, there are many traits relating cognitive models of agency to A&A. In fact, from a cognitive perspective, artifacts can be viewed as external entities that agents may exploit to improve their repertoire of actions. For knowledge representation in particular, *cognitive artifacts* – defined by Norman as those artifacts that maintain, display or operate upon information in order to serve a representational function [7] – are essential tools for reasoning, heavily affecting human behavior. Once an agent uses an artifact, he is both relying on its functionalities and delegating part of his purposive activities on artifact functioning. In this view, artifacts play the role of suitable services that agent can exploit to externalise activities thus easing their computational burdens [4]. An additional aspect relating cognitive agents to artifacts concerns the nature of open systems, exacting agents with the ability to discover information which is relevant for achieving their goals. These aspects heavily concern the ability to learn and co-use environmental and distributed resources in terms of subjective experiences and mental states.

The aim of this work is to enrich the abstraction tools in MAS design so as to provide alternatives from communication and cooperation strategies so far based solely on message passing and ACL, towards a view in which agents live in shared (distributed) work environments, which are part of their habitat, composed not only by other agents but also by dynamic sets of artifacts, created/scrutinized/disposed by agents themselves. In particular we discuss the abilities of agents to exploit artifacts with representational functionalities, i.e. cognitive artifacts. This, we guess, has manifold benefits, in particular in making it more effective to organise and share distributed knowledge, reasoning about distributed work, retrieving information, finally improving the coordination/cooperation typical of complex activities.

First we discuss the role played by cognitive artifacts upon the A&A model and CArtAgO technology (Section 2). Then, to provide some evidence of the overall model, we describe an experiment in the RoomsWorld scenario (Section 3), (Section 4) where specific kind of cognitive artifacts, i.e. *log*s, are exploited to support the distributed work and interaction of a team of goal-oriented agents. CArtAgO and *Jason* [1] are exploited as integrated technologies [11] to implement respectively the artifact-based environment and the cognitive agents.

2   Artifacts in Multi-Agent Systems

The notion of artifact in MAS has been introduced with the Agents & Artifacts A&A conceptual model as first-class abstraction along with agents when modeling, designing, and developing MAS [8]. A&A promotes the notion of artifacts and resources/tools playing a pivotal (mediation) role in coping with the scaling up of complexity, in particular when social interactions and complex activities are concerned [9]. In cognitive theories on human behavior, the use of tools can be considered along with an evolutionary accumulation and transmission of social knowledge, habits, memes, which not only influence the nature of external behavior, but also the mental functioning of individuals.

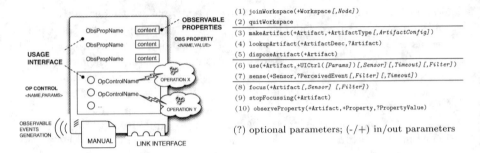

```
(1)  joinWorkspace(+Workspace [,Node])
(2)  quitWorkspace
(3)  makeArtifact(+Artifact,+ArtifactType [,ArtifactConfig])
(4)  lookupArtifact(+ArtifactDesc,?Artifact)
(5)  disposeArtifact(+Artifact)
(6)  use(+Artifact,+UICtrl([Params]) [,Sensor] [,Timeout] [,Filter])
(7)  sense(+Sensor,?PerceivedEvent [,Filter] [,Timeout])
(8)  focus(+Artifact [,Sensor] [,Filter])
(9)  stopFocussing(+Artifact)
(10) observeProperty(+Artifact,+Property,?PropertyValue)

(?) optional parameters; (-/+) in/out parameters
```

Fig. 1. *(Left)* An abstract representation of an artifact, with in evidence the usage interface, with operation controls and observable properties, the manual and the link interface, used to link together artifacts. *(Right)* Basic set of actions integrating agent platforms *(Jason* in this case) and CArtAgO.

By analogy, the basic idea of A&A is to define a notion of *artifact* in MAS as first-class computational entities, representing resources and tools that agent can exploit to perform their tasks, in particular social ones. Then, according to the A&A model, designers have two complementary abstraction tools to define their MAS. Differently from agents, which can be described as autonomous, pro- active (goal-oriented) entities encapsulating the control of their behaviour, artifacts are a non-autonomous and *function-oriented* computational objects, i.e., encapsulating some kind of service (function) to be exploited by agents.

In this view, the interaction between agents and artifacts is shaped on two fundamental activities played by agents in their environment: *action* and *perception*. In this view, agents use artifacts through their *usage interface*, which is a control interface, and gets their output by perceiving observable events generated with artifact functioning and observable properties which are part of the artifact state (Fig. 1 Left). Besides the notion artifact, A&A introduces the notion of *workspace* as logic container of agents and artifacts. Workspaces can be used to structure the overall sets of entities, defining a topology of MAS environment. In this view a MAS is conceived as a set of (distributed) workspaces, containing dynamic sets of agents working together by both directly communicating and sharing/co-using artifacts (Fig. 2 Right).

2.1   The Role of Cognitive Artifacts

As remarked in [7], cognitive artifacts shape the way in which human beeings pursue their activities interacting with their surroundings. Examples of cognitive artifacts used by humans are, for instance, activity lists or logs – useful to maintain and manage a shared list of task to do – and white-boards – to maintain shared knowledge and provide suitable operation to access and update it. Besides the specific functionalities, in [7] Norman remarks the benefits of cognitive artifacts under two different perspectives. From an individual point of view, artifacts basically make it possible to enhance agent cognitive capabilities, either by saving internal computational load or being merely memory enhancer, for instance creating observable cues in order to highlight relevant information. Besides, agents are not forced to contemporaneously communicate by means of message passing but they can asynchronously interact according to emerging dynamics (*situated cognition*). In addition, by changing the actions required for achieving a goal, artifacts change the way a task gets done. In so doing cognitive artifacts can help individual agents to eliminate decision points, creating more rational decision alternatives and reducing the *fan-out* of a decision tree [5].

From a system perspective, cognitive artifacts can be useful to: *(i)* distribute actions across time (*pre-computation*), by allowing to store information and enable information processing that can be performed even before the actual operation usage; *(ii)* distribute the actions across time and space (*distributed cognition*), asynchronously distributing tasks across agents and tools, allowing social coordination and shared activities.

Analogously to the Norman definition, we refer to *cognitive artifacts* for MAS those computational entities that are designed to maintain, make it observable, or operate information in order to serve a representational function for agents with respect to their environment and the work inside it. From an agent viewpoint, artifacts can be cognitively used once their representational content can be mapped and translated into reasoning processes. Relations from cognitive agents to cognitive artifacts are, at least, bidirectional. On the one side artifact interface descriptions can be matched with agents epistemic (beliefs) and motivational (goals) mental states. On the other side, the external events coming from a cognitive artifact can be integrated at an architectural level by automatically promoting such events as "relevant" signals to be addressed to the reasoning processes.

Then, by adopting a functional view, from an agent's perspective artifacts can be viewed as devices playing the role of targets for purposive activities, thus serving a twofold function: *doxastic* and *operational* [10]. On the one side, artifacts play an operational (purposive) function according to which goals can be achieved by means of operations which have been defined within artifact control interface. To this end, the A&A meta-model defines a series of general properties by which artifact functionalities are defined in terms of *operations*, which can be triggered by agents via artifact's *usage interface* (Fig. 1 Left). On the other side, artifacts play a doxastic (epistemic) function, being their representational and observable contents as external informational structures exploitable by agents.

2.2   Cognitive Agents Using Artifacts in CArtAgO

On the basis of the operational and doxastic functions of artifacts, A&A interactions can be based on agent's native abilities of action and perception. More formally, an agent AG *rely* on an artifact AR for a given goal G_i, according to a set of plans P_{ji} if:

1. AG has G_i in its set of goals;
2. AG is not autonomous for G_i, i.e. lacks at least one of the *resources* (*actions/information*) necessary to achieve G_i;

3. There is a plan $p_{ji} \in P_{ji}$ that achieves G_i where at least one resource (action/information) used in this plan is in AR's set of available operations or AR's observable state.

Analogously to usage interface of artifacts in the real world, an artifact usage interface is composed by a set of *operation controls* that agents can use to trigger and control operation execution. Each operation control is identified by a label (typically equal to the operation name to be triggered) and a list of input parameters. Operations are the basic unit upon which artifact functionality is structured. The execution of an operation can result both in changes in the artifact's inner (i.e. non-observable) state, and in the generation of a stream of *observable events* that can be perceived by agents that are using or simply observing the artifact. Besides, to support the doxastic function, the usage interface might contain also a set of *observable properties*, i.e. properties whose dynamic values can be observed by agents without necessarily acting upon operations.

Whereas A&A provide a conceptual underpinning on the agents and artifacts approach, CArtAgO [11] puts in practice the enabling technology to realise artifact-based MAS. CArtAgO offers a suitable integration technology supporting the implementation and the deployment of distributed MAS based on the A&A model, where agents dwelling in different agent platforms can autonomously join and work together in distributed and shared workspaces. It is worth noting that the platform promotes the deployment of open systems by allowing artifact-based environments and workspaces to persist – supported by the run time – even beyond the presence of agents. In addition CArtAgO resolves low-level issues at a platform level, for instance ruling interactions (i.e. action, perception, operation execution) and managing conflcts on resources (i.e. cuncurrency on shared artifacts, distribution across several nodes). Current CArtAgO distribution offers bridge mechanisms to be integrated by some well-known agent platforms (namely *Jason, Jadex,* 2APL)[2].

To allow cognitive agents to play in CArtAgO environments we consider basic sensory-motor aspects in a cognitive system fashion. On the one hand agents have been equipped with mechanisms to interact with artifacts (*effectors*); on the other hand they also have been enabled to perceive events generated by artifacts during their operations (*sensors*). In this view, the integration approach has been realized at the language level, i.e. the set of artifact-related actions have been added to the repertoire of natively available actions. Therefore, the bridge mechanism introduced the notion of *agent body* as that part of an agent conceptually belonging to a workspace (once the agent is inside it) and containing those sensory-motor capabilities to interact with artifacts. Fig. 1 (Right) shows basic actions integrated in the body allowing agents to interact with CArtAgO artifact-based environments. In particular, the described actions make it possible for an agent to: join, move, and leave workspaces (1-3); use an artifact by acting on its control interface and perceive events generated by artifacts (4-7); observe

---

[2] CArtAgO is available as an open source technology at:
    `http://cartago.sourceforge.net`. A detailed description of the platform – including aspects not essential for this work – can be found in CArtAgO site.

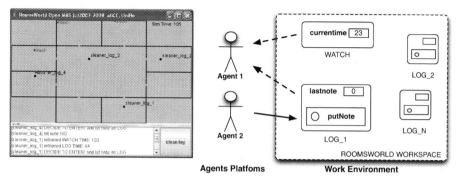

Fig. 2. (*Left*) The RoomsWorld scenario. Agents are engaged in cleaning thrash (red items) spread over eight rooms relying on artifact based facilities of watch and logs. (*Right*) An overview of the RoomsWorld scenario: agents perceiving and acting with artifacts distributed in CArtAgO workspaces.

artifact properties (8). A pseudo-code first-order logic-like notation is adopted for the syntax, while semantics is described informally. use is the basic action for triggering the execution of an operation, specifying operation control name and parameters, and optionally a sensor name. Sensors are conceived as a kind of body's *perceptual memory* to keep track of observable events generated by artifacts, possibly applying filters and specific kinds of "buffering" policies. If no sensor is specified, all the observable events generated by the artifact executing the operation are signalled as internal events to the agent. Otherwise, events collected in sensors can be retrieved by using the sense primitive, specifying a template (which functions as a filter) and possibly a timeout, indicating how long the current course of action can be suspended waiting for the event to be detected by the sensor. The basic support for artifact observation is provided by focus action, which makes it possible to continuously be aware of the observable events generated by the specified artifact. Finally, observeProperty is given to observe and get the value of a given observable property. In this case no sensors or percepts are involved: the value is directly bound to the variable specified within the basic action. It's worth remarking that, differently from use, focus and observeProperty do not cause the execution of any operation or computational work by the (observed) artifact.

To provide a practical example showing cognitive artifacts in practice, in the next sections we discuss the RoomsWorld scenario, where teams of goal-oriented agents exploit cognitive artifacts to perform their cooperative tasks.

3   Cognitive Artifacts in Practice

Built with CArtAgO technology, the RoomsWorld scenario realises an open system where heterogeneous agents have the possibility to join, test and interact with an artifact-based workspace. The workspace is composed by a number of virtual rooms separated by walls and doors (see Fig. 2 Left). Once a room is entered, agents should achieve the goal to find and clean trash items which may appear in the rooms with arbitrary frequencies. It is worth noting that, given its structure, the global environment is partially observable. Rooms are visitable one at a time and, to locate trash items, an agent has to enter and then perform some epistemic action. Once they get the percept of a trash, agents can reach its location and adopt a "clean" goal. For convenience, a global *environment* artifact is used to hold physical consistence to the simulation and supporting agents in their perceptive processes. The environment artifact is used here just as a convenient way to represent and implement in the simulation the physical environment: for instance, the location of the trash items are provided in the form of symbolic percepts which are assumed to contain the symbolic location of the trash.

To investigate effectiveness of artifact mediated interactions we engaged agents pursuing different strategies for achieving their goals in experiment series. The "normal cleaners" simply look for trash exploring the rooms based on personal experience. Agents in this case act as if they were alone in the environment, by autonomously updating relevant information without any co-operation by other member of the team. The second team of agents adopt a message based strategy: "Messenger cleaners" exploit explicit messages to coordinate themselves. Once a given room has been cleaned, a messenger shares this information by broadcasting a message to the overall society, thus indicating the room identifier and the exact time at which the clean activity has been finished. For simplicity, messengers have been implemented using the same agent platform (*Jason* in this case), thus exploiting their native infrastructure to process message passing.

To support the third strategy, a set of particular cognitive artifacts are deployed in the workspace. In this case agents have the possibility to purposively use special kind of artifacts which are physically placed at the entrance of each room. Such artifacts function as *logs* to supply agents with additional information which is relevant for accomplishing their work. For each Log, the @OPERATION void putNote(Object ts) allows agents to store the time-stamp at which a given clean activity has been performed (Table 1). Accordingly Logs expose @OBSPROPERTY int lastnote as an observable property, namely the last registered time-stamp, by which agents can infer the last performed clean activity for that room. Logs make it possible uncoupled interaction between who wants to share a given information (informing agent) and who may exploit this information (reading agent), thus allowing agents to be acknowledged on those particular information which is relevant to achieve the current tasks. In addition, agents also exploit a *watch* artifact, functioning as a timer which provides them

Table 1. CArtAgO implementation for the Log artifact

```
public class Log extends Artifact {         @OPERATION void putNote( Object ts ){
    private LinkedList<String> notes;            notes.addLast(timestamp.toString());
                                                 lastnote =
    @OPERATION void init(String n, location l){      ((Integer)timestamp).intValue();
      name=n; loc= l;                          }
      notes = new LinkedList<String>();     }
      defineObsProperty("lastnote", 0);
    }
}
```

with a symbolic record of the ongoing simulated time. Log strategy assumes that, before entering a given a room, agents retrieve the actual time from the watch and then store a time record in the related log. In the next section we detail the model for *Jason* agents adopting the log strategy.

3.1 Agent's Knowledge Model

Our proposed agent model adopts a systematic design structure for their beliefs. In particular agent's belief base can be logically structured in two different (sub)sets. The first set of beliefs indicates symbolic references to the generic objects of the simulation (i.e. system entities, workspaces, identifiers, registered CArtAgO sensors etc.). Belonging to this class, a nRooms(Nr) belief is updated once the agent joins the environment, Nr indicating the current number of rooms. Besides, agent's cognitive process relies on a second class of beliefs, taking into account the particular relation, in cognitive agents, between *relevant* information and goal processing. Two groups of beliefs belong to this class. The former belief set is composed by facts on the form cleaned(N,T) containing the state of the various rooms (N indicating a room identifier between 1 and Nr, T containing the time-stamp at which this information has been retrieved). The latter belief set is related to the room the agent has committed to clean (targetRoom(N)). We consider these beliefs as "salient" facts, namely the information which is *relevant* for agents for pursuing their goals and thus for selecting and committing the intention to clean a given room. The notion of relevance concerns here those information required to agents for ruling over deliberation and means-end reasoning. As in [3] we refer to this class of relevant beliefs as *goal-supporting beliefs*. Differently from the first defined class of beliefs, goal-supporting beliefs are dynamic ones: during agents' purposive activities they need to be updated in order to be consistent with world changes. Notice that information about the state of each room can be considered certain only at a given time t, afterwards the activities of other agents (or the rise of new trash items) would have modified the room state. As explained in section Section 4, we will exploit the number of belief update operations (pursued by agents upon these relevant facts) as a concrete measure for the computational load addressed by the cognitive processes.

3.2 Cleaner Agent Using Logs

For simplicity we here describe a cutout of the code for the Cleaner agent implementing the log strategy. Agents are implemented in AgentSpeak using the *Jason* platform [1] [3].

The initial goal join allows the agent to register himself to the RoomsWorld workspace and to join the environment artifact. The agent here uses the getNRooms operation retrieving the number of rooms n_rooms(Nr) where Nr unifies with the number of rooms signalled by the environment.

---

[3] AgentSpeak is an agent programming language for goal-oriented, BDI agents. The entire code of RoomsWorld experiment, along with agents implementation using alternative platforms, is available on CArtAgO web site.

```
+!join : myNameID(N) & mySensor(S);
<- cartago.joinWorkspace("RoomsWorld","localhost:4010");  +!explore : nRooms(N) & mySensor(S);
        !locate_artifacts;                                 <- -targetLog(_);
        ?artifactBel(environment, Env);                        roomsworld.randomInt(N,Rid);
        cartago.use(Env, join(N), S);                          +targetLog(Rid+1);
        cartago.sense(S, joined(Loc));                         ?artifactBel(watch, IDWatch);
        cartago.use(Env, getNRooms, S);                        cartago.observeProperty(IDWatch,
        cartago.sense(S, n_rooms(Nr));                                              currentime(Wt) );
        +nRooms(Nr);                                           !analizeBel(X+1,Wt).
        !explore.
```

Notice that a particular plan (`locate_artifacts`) is then called to retrieve and store the identifiers belonging to the artifacts that will be used during the task. Once retrieved through the use of `cartago.lookupArtifact` operation, the artifacts identifiers are stored in a belief set in the form `+artifactBel(artifactName, artifactId)` providing a knowledge repository associating each artifact name to the corresponding ID. Notice that those beliefs are not considered goal supporting beliefs, being static references to locate system resources more than information required to deliberate and make decisions.

The following `explore` goal is the starting point for agent's purposive activity, by which the agent randomly select a room `Rid` and prepare a new intention to explore in it. The action `cartago.use(W, whatTime, S)` executed upon the *watch* artifact allows the agent to retrieve the actual time: the agent observes the watch artifact, getting the content of the `time(Wt)` percept (W_T unifies with the percept content signalled by the watch)[4].

To filter out worth intentions and thus prevent exploring places which have been recently visited, the randomly selected room is then compared with the relevant belief base that refers to the acknowledged state of the rooms. So far, if the difference between the remembered time and the time at which the target room has been previously cleaned is greater than a given threshold D, the agent "commits" to the intention to go toward the room and then try to use the related log (i.e. by updating relevant belief `targetRoom(N)`). Otherwise, the agent abandons the current plan and readopt the explore goal to generate a different intention. Notice here that the agent is using his *relevant* information to make a decision about his next course of actions.

```
+!analizeBel(N,Wt) : not cleaned(N,Lt) | (cleaned(N,Lt) & day(D) & (D< T-Wt))
    <- -cleaned(N,_); -targetRoom(_); +targetRoom(N); !observeLog(N).
+!analizeBel(N,Wt) <- !explore.
```

So far, the agent observes the log artifact corresponding to the selected room and reads the last recorded time-stamp. This is done by executing the `observeProperty` action and perceiving the `lastnote` value, carrying to the agent the relevant information about the last time-stamp. We here focus on an interesting difference with respect to the strategies employed by other agent teams. Whereas normal agents have to autonomously update the knowledge about the problem domain and messenger agents have to continuously process

---

[4] The `cartago.*` actions refer to CArtAgO basic primitives (see Section 2.2), while `roomsworld.*` refer to the library of internal actions defined to operate within the RoomsWorld scenario.

incoming messages, here the log agent finds information by simply observing the log artifact on which he is interested. Relevant information left by some other agent at the end of a previous cleaning activity has been stored and collected within the artifact that makes it available for the overall society. This approach directly enables uncoupled interactions – mediated by the log – among different agents during their practical behavior.

```
+!observeLog(N) : targetCh(N) & myRoom(MR)
   <- roomsworld.goTo(MR, log(N));          cartago.observeProperty(IDLog, lastnote(LogT) );
      -targetCh(N);                         cartago.observeProperty(IDWatch, currentime(Wt) );
[ ... retrieve IDLog and IDWatch ... ]      !decide(N, Wt, LogT).
```

Once the agent has observed both information about log's last note Log_T and the information about the current time W_T provided by the watch, he can *decide* what to do next. Notice that the agent is here deciding upon a situated information (the actual state retrieved from the log corresponds to the last known state of the room), being this information (in the case of team activities) possibly updated by some other agent of the group. If the difference between the actual time W_T and the time-stamp retrieved in the log Log_T is greater than a given threshold (say, a day's length day(D)), the agent maintains his intention of entering the room N, otherwise he reconsiders his intentions and adopt a new **explore** goal to select a new room.

```
+!decide(N, CurrentTime, LastTime)
  : (day(D) & (D< CurrentTime-LastTime))
    | (LastTime=0)                          +!decide(N, CurrentTime, LastTime)
  <- !log("DECIDE TO ENTER! Put a note on Log");   <- !log("RECONSIDERED INTENTIONS:EXPLORING!") ;
     [ ... retrieve sensor S and Log name LogN ...]    -+cleaned(N,LastTime);
     cartago.lookupArtifact(LogN, IDLog);              -targetRoom(_);
     cartago.use(IDLog, putNote(CurrentTime), S);      !explore.
     !go(N).
```

It is worth noting that, within the decided plan, the agent updates either the state of his belief or the state of the log. Indeed once the agent has decided to enter the room, he puts a note on the log, either anticipating the information about that room to be cleaned or preventing follower agents to waste their resources going to explore the same room. Otherwise the agent needs to update the relevant beliefs and thus reconsider his intentions: in this case room N will be stored as 'cleaned' at time LastTime through the belief update +cleaned(N,LastTime). The following action go(N) allows the agent to enter the selected room N. After having updated the beliefs about target room and actual room, it's time for the agent to search for trash items: roomsworld.epistemicAction(N) is an internal action used to perceive the objects situated in the room. The epistemic action, in turn, encapsulates the use of the environment artifact and simulates agent's perceptive activities transforming volatile percepts (coming from CArtAgO sensors) into agent beliefs.

```
+!go(N): not myRoom(N)                      -+myRoom(N);
   <- .print("enter in room ", N);          roomsworld.epistemicAction(N);
      ?myRoom(MR); roomsworld.goTo(MR, room(N));  !clean(N).
      -targetRoom(_);                      +!go(N) <- !explore.
```

If trash is found in the room, the epistemic action adds beliefs in the form trash(N,X,Y) where N is the room ID, and X, Y are the coordinates of the discovered trash item. Once the agent has located trash items (if any), he can reach them with the action `roomsworld.goTo(X,Y)` and clean exploiting the `cartago.use(E, clean(Na, X, Y), S)` operation upon the environment E. Accordingly, the agent updates the beliefs referring to the state of the room. Otherwise, if the epistemic action has returned no items, the agent drops the goal, reconsider the current intention exploring another room.

Finally, the action to clean the trash item is realized calling the `cartago.use(Env, clean(Na, X, Y), S)` operation upon the `Env` artifact. The agent indicates his own name and the coordinates of the trash to clean (which is then returned as a percept). It is worth noting that the `clean/3` operation may fail, due to the non-determinism of the environment or due to the execution of concurrent cleaning actions performed by two or more agents on the same trash item at the same time. To verify the outcome of the clean operation the agent indicates a sensor S, which is used to collect the events signalled by the log.

4   RoomsWorld Experiment

To evaluate the different interaction strategies for cleaner agents in RoomsWorld described in Section 3, we ran sequences of trials measuring performances for different teams belonging to the different strategies. RoomsWorld was set to 8 rooms and the total amount of trash items contemporaneously present is limited to 4, whilst no more than one trash is generated for each room (Fig. 2). Each reported experiment consists of 14 repeated runs for each team, using different randomly generated initial conditions and distribution of trash items.

The payoffs of the three strategies are based on specific metrics taking into account the computational and behavioral effectivenes for the various agents. What we are interested in is a quantitative evaluation of the balance between computational costs spent by agents to update their beliefs and the absolute performances in terms of achieved goals (i.e., agent's score). This allows us to give an account about the agents reasoning processes in terms of computational load. For each trial we defined team's *goal effectiveness* in terms of cleaned items (i.e. agent's score) and in function of elapsed time. Besides, for each agent typology, we focused on the particular belief set which is involved in updating relevant information used to support reasoning and thus to perform the task. In more detail, we define the *belief change cost* as the aggregate of belief change operations performed by all the agents of the team upon their relevant beliefs (i.e. the sum of all modifications of agent's relevant belief base during all the trials). Given the structure of relevant belief set as it has been defined in Subsection 3.2, two kinds of updates entail an increase of belief change cost:

1. The state of a given room N may be known by agents as cleaned at a certain instant of time *Time* through the addition of relevant `+cleaned(N,Time)` facts. Once an agent has become aware of the new state of a room N (either because of a message sent by another agent, or because the agent has retrieved some log report, or because he has autonomously perceived the

state of a room) he immediately updates the relative facts. According to the definition, revising the relevant belief base with a new `cleaned(N,T)` fact entails a *belief change cost* of +1.

2. An additional cost is accounted by updating the belief on `targetRoom(N)`. In so doing, the agent reconsiders his intention selecting an alternative course of action. In turn, no more applicable plans will match the context-conditions about the room identifier. This entails the agent to reconsider his plans, selecting a new intention and exploring a new room.

The following cutout shows how the messenger agents are supposed to spend two units of *belief change cost*. Sender agents update the `cleaned(N,T)` belief once a room has been cleaned (*right* column) and must broadcast this information to the team; accordingly, a receiver agent updates his beliefs about the state of that room once a message is received (*left* column).

```
+!clean(N) : not trash(N,_,_)
  <- ?artifactBel(watch, W);
     cartago.observeProperty(W, sim_time(Wt));     +cleaned(room(N,T))[source(Ag)] : true
     .broadcast(tell, cleaned(room(N, Wt)));         <- -cleaned(room(N,_))[source(_)];
     -+cleaned(N,Wt);                                   -+cleaned(N,T);
     roomsworld.goTo(N, log(N));
     !explore.
```

For each team we also define the *cost effectiveness* ratio in terms of total amount of achieved tasks (agent's score) divided by the *belief change cost*. Namely, *cost effectiveness* represents the unit of achieved goals for each belief change.

   Conducted experiments considered the above defined metrics and used the threshold D to 12 and 6 units of simulated time respectively for teams of two and four agents. Because of the random distributions of items, the previously defined metrics present a fluctuating course before converging. Hence the length of each individual trial has been set sufficiently large to become stable and, after the analysis of the courses of experiments, it has been set to 250 units of simulated time. Fig. 3 and Fig. 4 show the performance of the different teams by averaging progresses of their effectiveness (on the left) and their cost ratio (on the right). In particular Fig. 3 refers for teams composed by two agents, while Fig. 4 to teams of four agents.

4.1   Experiment Discussion

Experiment results put in evidence the role of log artifacts functioning as cognitive resources for agents (a belief base enhancer in this case), clearly enabling agents to ease their epistemic activities. Anyhow, on the basis of the defined metrics, the results show some noticeable payoff for the various strategies. In the case of teams composed by two agents, messenger agents attain the best performance in terms of *goal effectiveness* (Fig. 3, left). Even if *goal effectiveness* for messenger and log agents approximatively converge to the same value (both reach a goal effectiveness of about 0.13 cleaned items on each elapsed time unit), messenger seems to tackle a shorter transitory phase. This evidence clearly comes from the fact that messengers can suddenly achieve and maintain an updated knowledge of the global states of the various rooms. Indeed, on

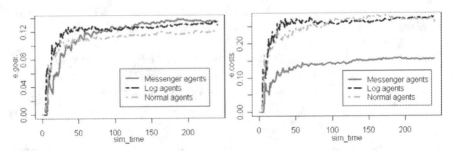

Fig. 3. Experiment 1: teams are composed by two agents using the same interaction strategy. Agents performances are measured in terms of *goal effectiveness (left)* referred to the amount of achieved goals, and *cost effectiveness (right)* referred to the computational load related to the update processing of relevant beliefs.

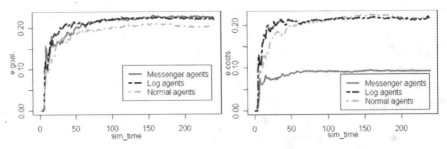

Fig. 4. Experiment 2: teams are composed by 4 agents. As before, performances are measured in terms of *goal effectiveness (left)*, and *costs effectiveness (right)*.

each achieved goal, message exchange allows agents to handle a more complete knowledge of environments. By augmenting the team members the global performance in terms of achieved goals for the various teams reach a higher value and messenger superiority in terms of achieved goals becomes less evident (Fig. 4, left). In this case, societies of numerous agents are better prone to make relevant information available to the overall group. Accordingly, agents using artifacts are more effective to cooperatively use the logs to get a more up-to-date state of the rooms. Hence, for teams composed by four agents, the *goal effectiveness* for messenger and log agents approximatively converge to the same value (both the teams reach a global goal effectiveness of about 0.22 cleaned items on each elapsed time unit). Besides, the log agents outperform normal agents in all conditions: whereas normal agents waste time and resources randomly looking for trash in rooms that have just been cleaned, log agents are smarter in recognising rooms requiring services, thus better balancing their payoff between exploration and exploitation.

Performance result inverted when considering the *cost effectiveness*: in this case the winning team is the one composed by normal agents, due to the lower amount of belief update performed upon the relevant facts and to the lower frequency of intention reconsideration. As Fig. 3 (right) shows, for each internal belief update the teams composed by two normal agents achieve 0.25 goals (0.27 for logging agents and 0.15 for messenger agents). Similar results have been collected for the teams composed by 4 agents (Fig. 4, right), where messenger agents attain a very poor performance (whilst normal and log agents achieve 0.22 cleaned trash items for each belief update, messenger agents reach the value of 0.09, due to the significant increase of message passing).

Balancing the performances for each selected metrics in each environmental condition, we here enlighten the effectiveness offered by the communication strategy mediated by log artifacts. The advantages of exploiting distributed logs to externalise information which is relevant for coordinating team activities are more evident the more the team is numerous. Logs give a considerable contribute both in terms of propagation and synchronization of the information. Even more, logs provide a informational support in uncertain, transitory conditions (i.e. when agents need to adjust behavior given environment partial observability). Although broadcasting messages to the overall society allows agents to maintain an updated knowledge of the global environment, this requires agents to waste their computational resources to continually process messages which are not strictly relevant for the ongoing task. On the contrary, log agents locally exploit information concerning their *actual* purposes, fully exploiting logs as suitable belief base enhancer. Differently from other approaches exploiting mediated interactions as shared memories and blackboards, logs are here conceived with the aim to improve *situated cognition*: agents can exploit logs to attain only the local information which is relevant to achieve the actual goal. Furthermore, the information coming from distant rooms is not relevant for agents because does not affect their ongoing intentions. This allows agents to deal with situated information, and may not require agents to pay attention (nor spend computational resources) to information which is useless with respect of the ongoing tasks. It is worth remarking that logs store the information which remains available even beyond their use. This has a pivotal importance in the context of open systems, where different agents may asynchronously operate, with interleaved presence, in specific tasks. Agents using logs are deliberatively modifying their environment with the aim to coordinate with other agents of the society and let them know the actual state of affairs. In so doing, the global behavior of the society is governed by the *emerging contents* of the distributed logs, which are cooperatively updated by agents during their activities.

5   Conclusions and Related Works

In this paper we discussed the role that cognitive artifacts, as artifacts with specific representational functionalities, can play in multi-agent systems, analogously to the human case, in supporting efficiently the cooperation of intelligent agents, who can share information by cooperatively updating the state of the artifacts, thus externalizing belief revision activities and easing their computational burdens. A first important outcome in exploiting cognitive artifact is in promoting mediated interaction, concerning communication and knowledge transmission. Exploiting artifact can be viewed as a complimentary approach with respect to direct communication based on messages and ACL. For this purpose, cognitive artifacts provide a natural support for time and space uncoupled communication, being then particularly useful in loose coupled interactions and open systems, where agents need to cooperate without necessarily being in the same temporal and spatial context.

This aspect is in agreement with many other research works recently appeared in literature (see [14] for a survey), remarking the role that the environment could play in designing complex MAS, as a suitable place where to encapsulate functionalities and services useful for agent interaction, coordination and cooperation. Among these, few works are specifically about cognitive MAS, in particular about high-level environment models specifically conceived to support and promote goal-oriented behavior of cognitive agents, and related agent reasoning techniques. A first one is Brahms [12], a multi-agent programming language and platform to develop and simulate multi-agent models of human and machine behavior, based on a theory of work practice and situated cognition. Our approach shares many points with Brahms, starting from a common reference conceptual background based on conceptual frameworks such as Activity Theory. Differently from Brahms, our primary context is not modeling and simulation, but agent-oriented software development. A further work is GOLEM [2], that introduces a platform for modeling situated cognitive agents in distributed environments by declaratively describing the representation of the environment in a logic-based form. GOLEM models (physical) environments in terms of *containers* where agents based on the KGP model and *objects* are situated. Besides sharing the same modeling perspective, which can be traced back to our early works on artifact-based coordination in MAS [9], we here investigate the cognitive use of artifacts, focussing in particular on the role of cognitive artifacts.

References

1. Bordini, R., Hübner, J.: BDI agent programming in AgentSpeak using Jason. In: Toni, F., Torroni, P. (eds.) CLIMA 2005. LNCS (LNAI), vol. 3900, pp. 143–164. Springer, Heidelberg (2006)
2. Bromuri, S., Stathis, K.: Situating Cognitive Agents in GOLEM. In: Weyns, D., Brueckner, S.A., Demazeau, Y. (eds.) EEMMAS 2007. LNCS (LNAI), vol. 5049, pp. 115–134. Springer, Heidelberg (2008)
3. Castelfranchi, C., Paglieri, F.: The role of beliefs in goal dynamics: Prolegomena to a constructive theory of intentions. Synthese 155, 237–263 (2007)
4. Clark, A., Chalmers, D.: The extended mind. Analysis 58(1), 7–19 (1998)
5. Kirsh, D.: The intelligent use of space. Artif. Intell. 73(1-2), 31–68 (1995)
6. Nardi, B.A.: Context and Consciousness: Activity Theory and Human-Computer Interaction. MIT Press, Cambridge (1996)
7. Norman, D.: Cognitive artifacts. In: Designing interaction: Psychology at the human–computer interface. Cambridge University Press, New York (1991)
8. Omicini, A., Ricci, A., Viroli, M.: Artifacts in the A&A meta-model for multi-agent systems. Autonomous Agents and Multi-Agent Systems 17(3) (2008)
9. Omicini, A., Ricci, A., Viroli, M., Castelfranchi, C., Tummolini, L.: Coordination Artifacts: Environment-based Coordination for Intelligent Agents. In: Proceedings of AAMAS 2004, New York, USA, vol. 1, pp. 286–293 (2004)

10. Piunti, M., Ricci, A.: From Agents to Artifacts Back and Forth: Purposive and Doxastic use of Artifacts in MAS. In: Proc. of 6th European Workshop on Multi-Agent Systems (EUMAS 2008), Bath, UK (2008)
11. Ricci, A., Piunti, M., Acay, L.D., Bordini, R., Hubner, J., Dastani, M.: Integrating Artifact-Based Environments with Heterogeneous Agent-Programming Platforms. In: Proceedings of AAMAS 2008 (2008)
12. Sierhuis, M., Clancey, W.J.: Modeling and simulating work practice: A human-centered method for work systems design. IEEE Intelligent Systems 17(5) (2002)
13. Weyns, D., Omicini, A., Odell, J.: Environment as a First-class Abstraction in MAS. In: Autonomous Agents and Multi-Agent Systems [14], pp. 5–30
14. Weyns, D., Parunak, H.V.D.: Special issue on environments for multi-agent systems. Autonomous Agents and Multi-Agent Systems 14(1), 1–116 (2007)

Information-Based Argumentation

Carles Sierra[1] and John Debenham[2]

[1] Institut d'Investigació en Intel·ligència Artificial – IIIA,
Spanish Scientific Research Council, CSIC
08193 Bellaterra, Catalonia, Spain
sierra@iiia.csic.es
[2] University of Technology, Sydney, Australia
debenham@it.uts.edu.au

Abstract. Information-based argumentation aims to model the partner's reasoning apparatus to the extent that an agent can work with it to achieve outcomes that are mutually satisfactory and lay the foundation for continued interaction and perhaps lasting business relationships. Information-based agents take observations at face value, qualify them with a belief probability and build models solely on the basis of messages received. Using augmentative dialogue that describes *what* is *good* or *bad* about proposals, these agents observe such statements and aim to model the way their partners react, and then to generate dialogue that works in harmony with their partner's reasoning.

1   Introduction

This paper is in the area labelled: *information-based agency* [1]. An information-based agent has an identity, values, needs, plans and strategies all of which are expressed using a fixed ontology in probabilistic logic for internal representation and in an illocutionary language [2] for communication. All of the forgoing is represented in the agent's deliberative machinery.

In line with our "Information Principle" [2], an information-based agent makes no *a priori* assumptions about the states of the world or the other agents in it — represented in a world model inferred from the messages that it receives. These agents build up their models by comparing expectation with observation — in this way we have constructed general models of trust, honour and reliability in a single framework [1].

[2] describes a rhetorical argumentation framework that supports argumentative negotiation. It does this by taking into account: the relative information gain of a new utterance and the relative semantic distance between an utterance and the dialogue history. Then [3] considered the effect that argumentative dialogues have on the on-going *relationship* between a pair of negotiating agents. Neither of these contributions addressed the relationship between argumentative utterances or strategies for argumentation. In this paper we adress these two issues.

The basis of our approach differs from [4] who builds on the notion of one argument "attacking" another. With the exception of a logical 'attack', whether one argument attacks another or not will depend on the receiving agent's private circumstances that are unlikely to be fully articulated. Thus, the notion of attack is of little use to information-based agents that build their models on the contents of utterances. This paper considers

J.-J.Ch. Meyer and J.M. Broersen (Eds.): KRAMAS 2008, LNAI 5605, pp. 130–144, 2009.

how to *counter* the effect of the partner agent's arguments, and aims to lead a negotiation towards some desired outcome by persuasive argumentation.

This paper is based in rhetorical argumentation [5]. For example, suppose I am shopping for a new car and have cited "suitability for a family" as a criterion. The salesman says "This LandMonster is great value," and I reply "My grandmother could not climb into that." Classical argumentation may attempt to refute the matriarch's lack of gymnastic prowess or the car's inaccessibility. Taking a less confrontational and more constructively persuasive view we might note that this statement impacts negatively on the "suitability for a family" criterion, and attempt to counter that impact possibly with "It's been voted No 1 for children." Although a smarter response may look for an argument that is semantically closer: "The car's height ensures a very comfortable ride over rough terrain that is popular with old people."

Information-based agents build their world models using an expectation/observation framework; this includes a model of the negotiation partner's behaviour. Agents form an *a priori* expectation of the significance of every event that occurs, and when the effect of that event is finally observed they revise their expectations. The model of behaviour includes measures of: trust, honour, reliability, intimacy, balance and disposition — *disposition* attempts to model what the partner means which may not be what they say. These measures are summarised using: temporal criteria, the structure of the ontology, and the illocutionary category of observed utterances.

Our argumentation agent has to perform two key functions: to understand incoming utterances and to generate responses. The approach is founded on a model of contract acceptance that is described in Section 2. Section 3 details a scenario that provides the context for the discussion. Sections 4 and 5 consider the scenario from each side of the bargaining table. Reactive and proactive argumentation strategies are given in Section 6, and Section 7 concludes.

2   Contract Acceptance

No matter what interaction strategy an agent uses, and no matter whether the communication language is that of simple bargaining or rich argumentation, a negotiation agent will have to decide whether or not to sign each contract on the table. We will argue in Section 4 that the buyer will be uncertain of his preferences in our Scenario described in Section 3. If an agent's preferences are uncertain then it may not make sense to link the agent's criterion for contract acceptance to a strategy that aims to optimise its utility. Instead, we pose the more general question: "how certain am I that $\delta = (\phi, \varphi)$ is a good contract to sign?" — under realistic conditions this may be easy to estimate. $\mathbb{P}^t(\text{sign}(\alpha, \beta, \chi, \delta))$ estimates the certainty, expressed as a probability, that α should sign[1] proposal δ in satisfaction of her need χ, where in (ϕ, φ) ϕ is α's commitment and φ is β's. α will accept δ if: $\mathbb{P}^t(\text{sign}(\alpha, \beta, \chi, \delta)) > c$, for some level of certainty c.

To estimate $\mathbb{P}^t(\text{sign}(\alpha, \beta, \chi, \delta))$, α will be concerned about what will occur if contract δ is signed. If agent α receives a commitment from β, α will be interested in any

---

[1] A richer formulation is $\mathbb{P}^t(\text{eval}(\alpha, \beta, \chi, \delta) = e_i)$ where $\text{eval}(\cdot)$ is a function whose range is some descriptive evaluation space containing terms such as "unattractive in the long term".

variation between β's commitment, φ, and what is actually observed, as the enactment, φ'. We denote the relationship between commitment and enactment:

$$\mathbb{P}^t(\text{Observe}(\alpha, \varphi')|\text{Commit}(\beta, \alpha, \varphi))$$

simply as $\mathbb{P}^t(\varphi'|\varphi) \in \mathcal{M}^t$, and now α has to estimate her belief in the acceptability of each possible outcome $\delta' = (\phi', \varphi')$. Let $\mathbb{P}^t(\text{acc}(\alpha, \chi, \delta'))$ denote α's estimate of her belief that the outcome δ' will be acceptable in satisfaction of her need χ, then we have:

$$\mathbb{P}^t(\text{sign}(\alpha, \beta, \chi, \delta)) = f(\mathbb{P}^t(\delta'|\delta), \mathbb{P}^t(\text{acc}(\alpha, \chi, \delta'))) \tag{1}$$

for some function f;[2] if f is the arithmetic product then this expression is mathematical expectation. f may be more sensitive; for example, it may be defined to ensure that no contract is signed if there is a significant probability for a catastrophic outcome.

There is no prescriptive way in which α should define $\mathbb{P}^t(\text{acc}(\alpha, \chi, \delta'))$, it is a matter for applied artificial intelligence to capture the essence of what matters in the application. In any real application the following three components at least will be required. $\mathbb{P}^t(\text{satisfy}(\alpha, \chi, \delta'))$ represents α's belief that enactment δ' will satisfy her need χ. $\mathbb{P}^t(\text{obj}(\alpha, \delta'))$ represents α's belief that δ' is a fair deal against the open marketplace — it represents α's *objective* valuation. $\mathbb{P}^t(\text{sub}(\alpha, \chi, \delta'))$ represents α's belief that δ' is acceptable in her own terms taking account of her ability to meet her commitment ϕ [2] [1], and any way in which δ' has value to her personally[3] — it represents α's *subjective* valuation. That is:

$$\mathbb{P}^t(\text{acc}(\alpha, \chi, \delta')) = g(\mathbb{P}^t(\text{satisfy}(\alpha, \chi, \delta')), \mathbb{P}^t(\text{obj}(\alpha, \delta')), \mathbb{P}^t(\text{sub}(\alpha, \chi, \delta'))) \tag{2}$$

for some function g.

Suppose that an agent is able to estimate: $\mathbb{P}^t(\text{satisfy}(\alpha, \chi, \delta'))$, $\mathbb{P}^t(\text{obj}(\alpha, \delta'))$ and $\mathbb{P}^t(\text{sub}(\alpha, \chi, \delta'))$. The specification of the aggregating g function will then be a strictly subjective decision. A highly cautious agent may choose to define:

$$\mathbb{P}^t(\text{acc}(\alpha, \chi, \delta')) = \begin{cases} 1 & \text{if: } \mathbb{P}^t(\text{satisfy}(\alpha, \chi, \delta')) > \eta_1 \\ & \wedge\, \mathbb{P}^t(\text{obj}(\alpha, \delta')) > \eta_2 \,\wedge\, \mathbb{P}^t(\text{sub}(\alpha, \chi, \delta')) > \eta_3 \\ 0 & \text{otherwise.} \end{cases}$$

for some threshold constants η_i. Whereas an agent that was prepared to permit some propagation of confidence from one factor to compensate another could define:

$$\mathbb{P}^t(\text{acc}(\alpha, \chi, \delta')) = \mathbb{P}^t(\text{satisfy}(\alpha, \chi, \delta'))^{\eta_1} \times \mathbb{P}^t(\text{obj}(\alpha, \delta'))^{\eta_2} \times \mathbb{P}^t(\text{sub}(\alpha, \chi, \delta'))^{\eta_3}$$

where the η_i balance the influence of each factor.

The point of this is: if an agent aims to produce persuasive argumentative dialogue then in the absence of any specific information concerning the structure of g the agent

---

[2] β influences the equation in the sense that different βs yield different $\mathbb{P}^t(\delta'|\delta)$.

[3] For example, when buying a new digital camera, α may give a high subjective valuation to a camera that uses the same memory cards as her existing camera.

should ignore g and concentrate on the three categories: $\mathbb{P}^t(\text{satisfy}(\alpha, \chi, \delta'))$, $\mathbb{P}^t(\text{obj}(\alpha, \delta'))$ and $\mathbb{P}^t(\text{sub}(\alpha, \chi, \delta'))$.

So how then will α specify: $\mathbb{P}^t(\text{satisfy}(\alpha, \chi, \delta))$, $\mathbb{P}^t(\text{sub}(\alpha, \chi, \delta))$ and $\mathbb{P}^t(\text{obj}(\alpha, \delta))$? Of these three factors only $\mathbb{P}^t(\text{obj}(\alpha, \delta))$ has a clear meaning, but it may only be estimated if there is sufficient market data available. In the case of selling sardines this may well be so, but in the case of Google launching a take-over bid for Microsoft it will not[4]. Concerning $\mathbb{P}^t(\text{satisfy}(\alpha, \chi, \delta))$ and $\mathbb{P}^t(\text{sub}(\alpha, \chi, \delta))$ we assume that an agent will somehow assess each of these as some combination of the confidence levels across a set of privately-known *criteria*. For example, if I am buying a camera then I may be prepared to define:

$$\mathbb{P}^t(\text{satisfy}(\alpha, \chi, \delta)) = h(\mathbb{P}^t(\text{easy-to-use}(\alpha, \delta)), \mathbb{P}^t(\text{well-built}(\alpha, \delta))) \tag{3}$$

for some function h. Any attempt to model another agent's h function will be as difficult as modelling g above. *But*, it is perfectly reasonable to suggest that by observing my argumentative dialogue an agent could form a view as to which of these two criteria above was more important.

This paper considers how an agent may observe the argumentative dialogue with the aim of modelling, within each of the three basic factors, the partner's criteria and the relative importance of those criteria. In repeated dealings between two agents, this model may be strengthened when the objects of the successive negotiations are semantically close but not necessarily identical.

3 The Scenario

Rhetorical argumentation is freed from the rigour of classical argumentation and descriptions of it can take the form of "this is how it works here" and "this is how it works there" without describing a formal basis. We attempt to improve on this level of vagary by using a general scenario and describing the behaviour of our agents within it.

In a general retail scenario there is a seller agent, α, and a buyer, β. The items for sale are abstracted from: digital cameras, mobile phones, PDAs, smart video recorders, computer software, sewing machines and kitchen mixers. The features of an item are those that are typically listed on the last few pages of an instruction booklet. For example, a camera's features could include the various shutter speeds that it is capable of, the various aperture settings, the number of years of warranty, and so on — together the *features* describe the capabilities of the item. For the purpose of comparison with other items, β will consider a particular item as a typed Boolean vector over the (possible) features of each item available, this vector shows which feature is present. The *state* of an item is then specified by identifying which of the item's features are 'on'. For example, the state of a camera could be: 'ready' with aperture set to 'f8' and shutter speed set to '1 500'th of a second'. In this scenario an *offer* is a pair (supply of a particular item, supply of some money) being α's and β's commitments respectively.

β may wish to know how well an item performs certain tasks. Software agents are not naturally endowed with the range of sensory and motor functions to enable such

---

[4] In this example the subjective valuation will be highly complex.

an evaluation. We imagine that the seller agent has an associated tame human who will demonstrate how the various items perform particular tasks on request, but performs no other function. We also imagine that the buyer agent has an associated tame human who can observe what is demonstrated, articulates an evaluation of it that is passed to its own agent, but performs no other function.

To simplify our set up we assume that the seller, α, is β's only source of information about what tasks each item can perform, and, as we describe below, what sequence of actions are necessary to make an item perform certain tasks[5]. That is, our multiagent system consists only of $\{\alpha, \beta\}$, and the buyer is denied access to product reviews, but *does* have access to market pricing data. This restriction simplifies the interactions and focusses the discussion on the argumentation.

For example, if the item is a camera the buyer may wish to observe how to set the camera's states so that it may be used for 'point-and-shoot' photography. If the item is a sewing machine she may wish to see how to make a button hole on a piece of cloth. If the item is graphics software she may wish to see how to draw a polygon with a two-pixel red line and to colour the polygon's interior blue. These tasks will be achieved by enacting a process that causes the item to pass though a sequence of states that will be explained to β by α. So far our model consists of: features, states, sequences and tasks.

We assume that the object of the negotiation is clear where the object is an uninstantiated statement of what both agents jointly understand as the intended outcome — e.g. I wish to exchange a quantity of eggs of certain quality for cash. We assume that each agent is negotiating with the aim of satisfying some goal or need that is private knowledge. In determining whether a negotiation outcome is acceptable in satisfaction of a need we assume that an agent will blend the factors in our acceptance model described in Section 2. We assume that for each factor an agent will articulate a set of *criteria* that together determine whether the factor is acceptable. The criteria may include private information such as deadlines.

More formally, there is a set of feature names, $\mathcal{F}$, a set of item names, $\mathcal{I}$, a feature mapping: feature : $\mathcal{I} \rightarrow \times^n(\mathbb{B} : \mathcal{F})$ where there are n feature names, and $\mathbb{B}$ is a boolean variable that may be $\top$ or $\bot$. Each item name belongs to a unique concept — e.g.: "Nikon123 is-a camera". For any particular item name, ν, feature(ν) will be a typed Boolean vector indicating which features that item ν possesses. Let $\mathcal{F}_\nu$ be the set of n_ν features that item ν possesses. At any particular time t, the state of an item is a mapping: state^t : $\mathcal{I} \rightarrow \times^{n_\nu}(\mathbb{B} : \mathcal{F}_\nu)$ where the value $\top$ denotes that the corresponding feature of that item is 'on'. A *sequence* is an ordered set of states, (w_i), where successive states differ in one feature only being on and off. A sequence is normally seen as performing a *task* that are linked by the mapping: to-do : $\mathcal{T} \rightarrow 2^S$ where $\mathcal{T}$ is the set of tasks and S the set of all possible sequences — that is, there many be several sequences that perform a task. If a sequence is *performed* on an item then, with the assistance of a human, the agent rates how well it believes the sequence performs the associated task. The evaluation space, $\mathcal{E}$, could be $\{good, OK, bad\}$. A criterion is a predicate: criterion(ν), meaning that the item ν satisfies criterion 'criterion'. The set of criteria is

---

[5] In other words, the sort of information that is normally available in the item's Instruction Booklet — we assume that α conveys this information accurately.

$\mathcal{C}$. The argumentation requirements include (where $x \in V$, $c \in \mathcal{C}$, $v = \text{feature}(x)$, $y \in \mathcal{T}$, $z \in \mathcal{S}$, and $r \in \mathcal{R}$):

- "I need an x"
- "What sort of x do you need?"
- "I need an x that satisfies criterion c"
- "What features does x have?"
- "x has features v"
- "How do you make x do y"
- "The sequence z performed on x does y"
- "Perform sequence z on x"
- "If sequence z is performed on x then how would you rate that?"
- "I rate the sequence z as performed on x as r"

4   The Buyer Assesses a Contract

In this Section we consider how the buyer might use the general framework in Section 2 to assess a contract[6]. In general an agent will be concerned about the enactment of any contract signed as described in Equation 1. In the scenario described in Section 3, enactment is not an issue, and so we focus on Equation 2. To simplify things we ignore the subjective valuation factor. Before addressing the remaining two factors we argue that the buyer will not necessarily be preference aware.

Consider a human agent with a need for a new camera who goes to a trusted camera shop. If the agent is preference aware he will be able to place the twenty to fifty cameras on offer in order of preference. If is reasonable to suggest that a normal, intelligent human agent could not achieve this with any certainty, nor could he with confidence represent his uncertainty in his preferences as a probability distribution over his preferences. This lack of awareness of preferences may be partially due to lack of information about each camera. But, what could "perfect information" realistically mean in this example? Even if the purchaser could borrow all the cameras for a day and had access to trusted, skilled users of each camera even then we suggest that our human agent would still be unable to specify a preference order with confidence. The underlying reason being the size and complexity of the issue space required to describe all of the features of every camera on offer, and the level of subjective judgement required to relate combinations of those features to meaningful criteria.

In large issue spaces, in terms of which an agent is unable to specify a preference ordering, there is a useful special case when it is possible to specify preferences on each issue individually (e.g. "I prefer to pay less than more", "I prefer to have a feature on the camera to not having it"). In this case the agent is *individual preference aware*.

4.1   Assessing $\mathbb{P}^t(\text{satisfy}(\beta, \chi, \delta))$

First β must give meaning to $\mathbb{P}^t(\text{satisfy}(\beta, \chi, \delta))$ by defining suitable criteria and the way that the belief should be aggregated across those criteria. Suppose one of β's criteria is $\mathbb{P}^t(\text{ease-of-use}(\beta, \delta))$. The idea is that β will ask α to demonstrate how certain

---

[6] The seller will have little difficulty in deciding whether a contract is acceptable if he knows what the items cost.

tasks are performed, will observe the sequences that α performs, and will use those observations to revise this probability distribution until some clear verdict appears.

Suppose the information acquisition process is managed by a plan π. Let random variable X represent $\mathbb{P}^t$(ease-of-use$(\beta, \delta) = e_i$) where the e_i are values from an evaluation space that could be $\mathcal{E} =$ {fantastic, acceptable, just OK, shocking}. Then given a sequence s that was supposed to achieve task τ, suppose that β's tame human rates s as evidence for ease-of-use as $e \in \mathcal{E}$ with probability z. Suppose that β attaches a weighting $\mathbb{R}^t(\pi, \tau, s)$ to s, $0 < \mathbb{R} < 1$, which is β's estimate of the *significance* of the observation of sequence s within plan π as an indicator of the true value of X. For example, the on the basis of the observation alone β might rate ease-of-use as $e =$ acceptable with probability $z = 0.8$, and separately give a weighting of $\mathbb{R}^t(\pi, \tau, s) = 0.9$ to the sequence s as an indicator of ease-of-use. For an information-based agent each plan π has associated *update functions*, $J_\pi(\cdot)$, such that $J_\pi^X(s)$ is a set of linear constraints on the posterior distribution for X. In this example, the posterior value of 'acceptable' would simply be constrained to 0.8.

Denote the prior distribution $\mathbb{P}^t(X)$ by p, and let $p_{(s)}$ be the distribution with minimum relative entropy[7] with respect to p: $p_{(s)} = \arg\min_r \sum_j r_j \log \frac{r_j}{p_j}$ that satisfies the constraints $J_s^X(s)$. Then let $q_{(s)}$ be the distribution:

$$q_{(s)} = \mathbb{R}^t(\pi, \tau, s) \times p_{(s)} + (1 - \mathbb{R}^t(\pi, \tau, s)) \times p \qquad (4)$$

and then let:

$$\mathbb{P}^t(X_{(s)}) = \begin{cases} q_{(s)} & \text{if } q_{(s)} \text{ is more interesting than } p \\ p & \text{otherwise} \end{cases} \qquad (5)$$

A general measure of whether $q_{(s)}$ is more interesting than p is: $\mathbb{K}(q_{(s)}\|\mathbb{D}(X)) > \mathbb{K}(p\|\mathbb{D}(X))$, where $\mathbb{K}(x\|y) = \sum_j x_j \log \frac{x_j}{y_j}$ is the Kullback-Leibler distance between two probability distributions x and y, and $\mathbb{D}(X)$ is the expected distribution in the absence of any observations — $\mathbb{D}(X)$ could be the maximum entropy distribution. Finally, $\mathbb{P}^{t+1}(X) = \mathbb{P}^t(X_{(s)})$. This procedure deals with integrity decay, and with two probabilities: first, the probability z in the rating of the sequence s that was intended to achieve τ, and second β's weighting $\mathbb{R}^t(\pi, \tau, s)$ of the significance of τ as an indicator of the true value of X. Equation 5 is intended to prevent weak information from decreasing the certainty of $\mathbb{P}^{t+1}(X)$. For example if the current distribution is $(0.1, 0.7, 0.1, 0.1)$, indicating an "acceptable" rating, then weak evidence $\mathbb{P}(X =$ acceptable$) = 0.25$ is discarded.

Equation 4 simply adds in new evidence $p_{(s)}$ to p weighted with $\mathbb{R}^t(\pi, \tau, s)$. This is fairly crude, but the observations are unlikely to be independent and the idea is that π

---

[7] Given a probability distribution q, the *minimum relative entropy distribution* $p = (p_1, \ldots, p_I)$ subject to a set of J linear constraints $g = \{g_j(p) = a_j \cdot p - c_j = 0\}, j = 1, \ldots, J$ (that must include the constraint $\sum_i p_i - 1 = 0$) is: $p = \arg\min_r \sum_j r_j \log \frac{r_j}{q_j}$. This may be calculated by introducing Lagrange multipliers λ: $L(p, \lambda) = \sum_j p_j \log \frac{p_j}{q_j} + \lambda \cdot g$. Minimising L, $\{\frac{\partial L}{\partial \lambda_j} = g_j(p) = 0\}, j = 1, \ldots, J$ is the set of given constraints g, and a solution to $\frac{\partial L}{\partial p_i} = 0, i = 1, \ldots, I$ leads eventually to p. Entropy-based inference is a form of Bayesian inference that is convenient when the data is sparse [6] and encapsulates common-sense reasoning [7].

will specify a "fairly comprehensive" set of tasks aimed to determine $\mathbb{P}^t(X)$ to a level of certainty sufficient for Equation 2.

4.2   Assessing $\mathbb{P}^t(\mathrm{obj}(\alpha, \delta))$

$\mathbb{P}^t(\mathrm{obj}(\beta, \delta))$ estimates the belief that δ is acceptable in the open-market that β may observe in the scenario. Information-based agents model what they don't know with certainty as probability distributions. Suppose that X is a discrete random variable whose true value is the open-market value of an item. First, β should be able to bound X to an interval $(x_{\min}, x_{\max})$ — if this is all the evidence that β can muster then X will be the flat distribution (with maximum entropy) in this interval, and $\mathbb{P}^t(\mathrm{obj}(\beta, (\mathrm{item}, y))) = \sum_{x \geq y} \mathbb{P}(X = x)$. β may observe evidence, perhaps as observed sale prices for similar items, that enables him to revise particular values in the distribution for X. A method [2] similar to that described in Section 4.1 is used to derive the posterior distribution — it is not detailed here. An interesting aspect of this approach is that it works equally well when the valuation space has more than one dimension.

5   The Seller Models the Buyer

In this Section we consider how the seller might model the buyer's contract acceptance logic in an argumentative context. As in Section 4 we focus on Equation 2 and for reasons of economy concentrate on the factor: $\mathbb{P}^t(\mathrm{satisfy}(\alpha, \chi, \delta))$.

5.1   Modelling Contract Acceptance

Suppose that β has found an item that he wants to buy, α will be interested in how much he is prepared to pay. In a similar way to Section 4.2, α can interpret β's proposals as willingness to accept the offers proposed, and counter-offers as reluctance to accept the agent's prior offer — all of these interpretations being qualified with an epistemic belief probability. Entropy-based inference is then used to derive a complete probability distribution over the space of offers for a random variable that represents the partner's limit offers. This distribution is "the least biased estimate possible on the given information; i.e. it is maximally noncommittal with regard to missing information" [8]. If there are n-issues then the space of limit offers will be an $(n-1)$-dimensional surface through offer space. As described in [2], this method works well as long as the number of issues is not large and as long as the agent is aware of its partner's preferences along each dimension of the issue space.

5.2   Estimating β's Key Criteria

α's world model, $\mathcal{M}^t$, contains probability distributions that model the agent's belief in the world, including the state of β. In particular, for every criterion $c \in \mathcal{C}$ α associates a random variable C with probability mass function $\mathbb{P}^t(C = e_i)$.

   β may present information in the form of a high-level description of what is required; e.g. "I want a camera for every-day family use". In a sense this is nothing more than a criterion, but it does not fit comfortably within the terms of Equation 2 and may have implications for all of them. We call such a statement as the *object* of the negotiation. It is realistic to assume that the object is common knowledge to some degree. We assume that there is a structured section of the ontology that describes negotiation objects. We also assume that for each object there is are prior probabilities associated with each of a set of negotiation criteria represented in another section of the ontology. For example, the object "A camera for every-day family use" may associate the prior probability distribution $(0.6, 0.4, 0.0, 0.0)$ with "ease of use for point-and-shoot" in terms of the example evaluation space given above.

The distributions that relate object to criteria may be learned from prior experience. If $\mathbb{P}^t(C = e|O = o)$ is the prior distribution for criteria C over an evaluation space given that the object is o, then given evidence from a completed negotiation with object o we use the standard update procedure described in Section 4.1. For example, given evidence that α believes with probability p that $C = e_i$ in a negotiation with object o then $\mathbb{P}^{t+1}(C = e|O = o)$ is the result of applying the constraint $\mathbb{P}(C = e_i|O = o) = p$ with minimum relative entropy inference as described previously, where the result of the process is protected by Equation 5 to ensure that weak evidence does not override prior estimates.

In the absence of evidence of the form described above, the distributions, $\mathbb{P}^t(C = e|O = o)$, should gradually tend to ignorance. If a decay-limit distribution [2] is known they should tend to it otherwise they should tend to the maximum entropy distribution.

In our scenario, during a dialogue β will ask α to perform certain tasks that we assume are represented in a structured section of the ontology. Following the reasoning above, if α is asked to perform task τ then this may suggest prior beliefs of what β's criteria are. For example, suppose α is asked to demonstrate how to photograph a duck when the object is a "camera suitable for photographing wildlife". If the ontology relates ducks to water to some degree then α may believe that β rates the criterion "waterproof" as "I would prefer to have this" with some probability. Then if α is subsequently asked to to demonstrate how to photograph a duck when the object is a s' she may believe that "waterproof" is a criterion. Using the semantic-similarity-based method described in [2], this evidence would update the estimates for $\mathbb{P}^t(\text{Waterproof} = e|(O = o', T = \tau))$ in a way that is moderated by the semantic distance between s' and 'camera suitable for photographing wildlife'"

The discussion so far has not considered argumentative dialogue such as: "My grandmother could not climb into that (car)". This statement would presumably follow a request to demonstrate the task "how to get into the car" that the observer would rate against the criterion "suitable for an octogenarian" as "unacceptable". So this example at least can be accommodated in the framework as long as we can link the 'grandmother' with an 'octogenarian'. Given two strings of ontological concepts, S_1 and S_2, define their *similarity* as:

$$\Pi(S_1, S_2) = \frac{\sum_{x \in S_1, y \in S_2} \text{Sim}(x, y)}{||S_1|| \times ||S_2||} \qquad (6)$$

where $||S||$ is the number of concepts in S, and $\text{Sim}(\cdot)$ is measures semantic distance [9]. A semantically deeper analyses of text would be better, but we claim that this approach will link such the given statement with the criterion "suitable for an octogenarian" — particularly if the set of admissible criteria are known and agreed in advance.

5.3   Disposition: Shaping the Stance

Agent β's *disposition* is the underlying rationale that he has for a dialogue. α will be concerned with the confidence in α's beliefs of β's disposition as this will affect the certainty with which α believes she knows β's key criteria. Gauging disposition in human discourse is not easy, but is certainly not impossible. We form expectations about what will be said next; when those expectations are challenged we may well believe that there is a shift in the rationale.

The bargaining literature consistently advises (see for example [10]) that an agent should change its *stance* (one dimension of stance being the 'nice guy' / 'tough guy' axis) to prevent other agents from decrypting their private information, and so we should expect some sort of "smoke screen" surrounding any dialogue between competitive agents. It would be convenient to think of disposition as the mirror-image of stance, but what matters is the agent's confidence in its model of the partner. The problem is to differentiate between a partner that is skilfully preventing us from decrypting their private information, and a partner that has either had a fundamental change of heart or has changed his mind in a way that will significantly influence the set of contracts that he will agree to. The first of these is normal behaviour, and the second means that the models of the partner may well be inaccurate.

If an agent believes that her partner's disposition has altered then the entropy of her model of the partner should be increased — particularly beliefs concerning the key criteria should be relaxed to prevent the dialogue attempting to enter a "blind alley", and to permit the search for common ground to proceed on broader basis. The mechanics for achieving this are simple: if an agent believes that his partner's disposition has shifted then his certainty of belief in the structure of the model of the partner is decreased.

α's model of β's *disposition* is $D_C = \mathbb{P}^t(C = e | O = o)$ for *every* criterion in the ontology, where o is the object of the negotiation. α's confidence in β's disposition is the confidence he has in these distributions. Given a negotiation object o, confidence will be aggregated from $\mathbb{H}(C = e | O = o)$ for *every* criterion in the ontology. Then the

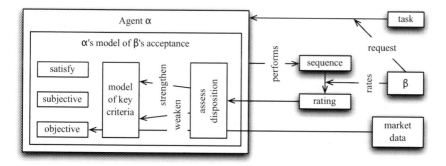

Fig. 1. The model of β's acceptance criteria that lies at the heart of the argumentation strategy

idea is that if in the negotiation for a camera "for family use" α is asked to demonstrate how to photograph a drop of water falling from a tap then this would presumably cause a dramatic difference between $\mathbb{P}^t(C = e|(O = $ "family use")) and $\mathbb{P}^t(C = e|(O = $ "family use", $O' = $ "photograph water drops")). This difference causes α to revise her belief in "family use", to revise the disposition towards distributions of higher entropy, and to approach the negotiation on a broader basis. A high-level diagram of α's model of β's acceptance criteria that includes disposition is shown in Figure 1.

6 Strategies

In this section we describe the components of an argumentation strategy starting with tools for valuing information revelation that are used to model the fairness of a negotiation dialogue. This work compares with [11], [12] and [13].

6.1 Information Revelation: Computing Counter Proposals

Everything that an agent communicates gives away information. The simple offer "you may purchase this wine for €3" may be intrepretd in a utilitarian sense (e.g. the profit that you could make by purchasing it), and as information (in terms of the reduction of your entropy or uncertainty in your beliefs about my limit price for the item). Information-based agents value information exchanged, and attempt to manage the associated costs and benefits.

Illocutionary categories and an *ontology* together form a framework in which the value of information exchanged can be categorised. The LOGIC framework for argumentative negotiation [3] is based on five illocutionary categories: Legitimacy of the arguments, Options i.e. deals that are acceptable, Goals i.e. motivation for the negotiation, Independence i.e: outside options, and Commitments that the agent has including its assets. In general, α has a set of illocutionary categories $\mathcal{Y}$ and a categorising function $\kappa : \mathcal{L} \rightarrow \mathcal{P}(\mathcal{Y})$. The power set, $\mathcal{P}(\mathcal{Y})$, is required as some utterances belong to multiple categories. For example, in the LOGIC framework the utterance "I will not pay more for a bottle of Beaujolais than the price that John charges" is categorised as both Option (what I will accept) and Independence (what I will do if this negotiation fails).

Then two central concepts describe relationships and dialogues between a pair of agents. These are *intimacy* — degree of closeness, and *balance* — degree of fairness. In this general model, the *intimacy* of α's relationship with β, A^t, measures the amount that α knows about β's private information and is represented as real numeric values over $\mathcal{G} = \mathcal{Y} \times V$.

Suppose α receives utterance u from β and that category $y \in \kappa(u)$. For any concept $x \in V$, define $\Delta(u, x) = \max_{x' \in concepts(u)} \mathrm{Sim}(x', x)$. Denote the value of A_i^t in position (y, x) by $A_{(y,x)}^t$ then:

$$A_{(y,x)}^t = \rho \times A_{(y,x)}^{t-1} + (1 - \rho) \times \mathbb{I}(u) \times \Delta(u, x)$$

for any x, where ρ is the discount rate, and $\mathbb{I}(u)$ is the *information*[8] in u. The *balance* of α's relationship with β_i, B^t, is the element by element numeric difference of A^t and α's estimate of β's intimacy on α.

We are particularly interested in the concept of intimacy in so far as it estimates what α knows about β's criteria, and about the certainty of α's estimates of the random variables $\{C_i\}$. We are interested in balance as a measure of the 'fairness' of the dialogue. If α shows β how to take a perfect photograph of a duck then it is reasonable to expect some information at least in return.

Moreover, α acts proactively to satisfy her needs — that are organised in a hierarchy[9] of *needs*, Ξ, and a function $\omega : \Xi \rightarrow \mathcal{P}(W)$ where W is the set of perceivable states, and $\omega(\chi)$ is the set of states that satisfy need $\chi \in \Xi$. Needs turn 'on' spontaneously, and in response to *triggers*. They turn 'off' because α believes they are satisfied. When a need fires, a plan is chosen to satisfy that need (we do not describe plans here). If α is to contemplate the future she will need some idea of her future needs — this is represented in her *needs model*: $\upsilon : T \rightarrow \times^{|\Xi|}[0,1]$ where T is time, and: $\upsilon(t) = (\chi_1^t, \ldots, \chi_{|\Xi|}^t)$ where $\chi_i^t = \mathbb{P}(\text{need } \chi_i \text{ fires at time } t)$.

Given the needs model, υ, α's *relationship model* (Relate$(\cdot)$) determines the target intimacy, A_i^{*t}, and target *balance*, B_i^{*t}, for each agent i in the known set of agents $Agents$. That is, $\{(A_i^{*t}, B_{*i}^t)\}_{i=1}^{|Agents|} = \text{Relate}(\upsilon, \mathbf{X}, \mathbf{Y}, \mathbf{Z})$ where, $\mathbf{X}_i$ is the trust model, $\mathbf{Y}_i$ is the honour model and $\mathbf{Z}_i$ is the reliability model as described in [2]. As noted before, the values for intimacy and balance are not simple numbers but are structured sets of values over $\mathcal{Y} \times V$.

When a need fires α first selects an agent β_i to negotiate with — the social model of trust, honour and reliability provide input to this decision, i.e. $\beta_i = \text{Select}(\chi, \mathbf{X}, \mathbf{Y}, \mathbf{Z})$. We assume that in her social model, α has medium-term intentions for the state of the relationship that she desires with each of the available agents — these intentions are represented as the target intimacy, A_i^{*t}, and target balance, B_i^{*t}, for each agent β_i. These medium-term intentions are then distilled into short-term targets for the intimacy, A_i^{**t}, and balance, B_i^{**t}, to be achieved in the current dialogue Ψ^t, i.e. $(A_i^{**t}, B_i^{**t}) = \text{Set}(\chi, A_i^{*t}, B_i^{*t})$. In particular, if the balance target, B_i^{**t}, is grossly exceeded by β failing to co-operate then it becomes a trigger for α to terminate the negotiation.

6.2 Computing Arguments

For an information-based agent, an incoming utterance is only of interest if it reduces the uncertainty (entropy) of the world model in some way. In information-based argumentation we are particularly interested in the effect that an argumentative utterance has in the world model including β's disposition, and α's estimate of β's assessment of current proposals in terms of its criteria.

Information-based argumentation attempts to counter the effect of the partner's arguments, in the simple negotiation protocol used here, an argumentative utterance, u,

---

[8] Information is measured in the Shannon sense, if at time t, α receives an utterance u that may alter this world model then the (Shannon) *information* in u with respect to the distributions in $\mathcal{M}^t$ is: $\mathbb{I}(u) = \mathbb{H}(\mathcal{M}^t) - \mathbb{H}(\mathcal{M}^{t+1})$.

[9] In the sense of the well-known Maslow hierarchy [14], where the satisfaction of needs that are lower in the hierarchy take precedence over the satisfaction of needs that are higher.

will either contain a justification of the proposal it accompanies, a rating and justification of one of α demonstration sequences, or a counter-justification of one of α's prior proposals or arguments. If u requests α to perform a task then u may modify β's disposition i.e. the set of conditional estimates of the form: $\mathbb{P}^t(C = e | O = o))$. If β rates and comments on the demonstration of a sequence then this affects α's estimate of β's likelihood to accept a contract as described in Equation 1 (this is concerned with *how* β will apply his criteria).

Suppose that u rates and comments on the performance of a sequence then that sequence will have been demonstrated in response to a request to perform a task. Given a task, τ, and a object, s, α may have estimates for $P^t(C = e | (O = o, \mathcal{T} = \tau))$ — if so then this suggests a link between the task and a set of one or more criteria C_u. The effect that u has on β's criteria (what ever they are) will be conveyed as the rating. In the spirit of the scenario, we assume that for every criterion and object pair (C, o) α has a supply of positive argumentative statements $\mathcal{L}_{(C,o)}$. Suppose α wishes to counter the negatively rated u with a positively rated u'. Let Ψ_u be the set of all arguments exchanged between α and β prior to u in the dialogue. Let $M_u \subseteq \mathcal{L}_{(C,o)}$ for any $C \in C_\mu$. Let $N_u \subseteq M_u$ such that $\forall x \in N_u$ and $\forall u' \in \Psi_u$, $\mathrm{Sim}*(concepts(x), concepts(u')) > \eta$ for some constant η. So N_u is a set of arguments all of which (a) have a positive effect on at least one criterion associated with the negative u, and (b) are at 'some distance' (determined by r) from arguments already exchanged. Then:

$$u' = \begin{cases} \arg\min_{u' \in N_u} \mathrm{Sim}*(concepts(u), concepts(u')) & \text{if } N_u \neq \emptyset \\ \arg\min_{u' \in M_u} \mathrm{Sim}*(concepts(u), concepts(u')) & \text{otherwise.} \end{cases}$$

So using only 'fresh' arguments, α prefers to choose a counter argument to u that is semantically close to u, and if that is not possible she chooses an argument that has some general positive effect on the criteria and may not have been used previously.

Suppose that u proposes a contract. α will either decide to accept it or to make a counter offer. We do not describe the bargaining process here, see [2].

6.3 All Together

If β_i communicates u then α responds with:

$$u' = Argue(u, \mathcal{M}^t, \Psi^t, A^{**t}, B^{**t}, C_u, N_u, M_u, D_u))$$

where:

- the *negotiation* mechanisms as explained in Section 6.1 sets parameters A^{**t}, B^{**t}) (see e.g. [3] for further details);
- the *argumentation* process determines the parameters N_u, M_u needed to generate the accompanying arguments to the proposal, see Section 6.2;
- the *criteria* modeling process determines the set of criteria C_u used by our opponent to assess the proposals, see Section 5.2; and,
- the *disposition* modeling sets the distributions D_u used to interpret the stance of the opponent, see Section 5.3.

The personality of the agent will be determined by the particular f chosen to select the answer to send. The study of concrete functions is subject of ongoing research as well as their application into a eProcurement setting.

7 Discussion

We have described an approach to argumentation that aims to:

- discover what the partner's key evaluative criteria are,
- model how the partner is evaluating his key criteria given some evidence,
- influence the partner's evaluation of his key criteria,
- influence the relative importance that the partner attaches to those criteria, and
- introduce new key criteria when it is strategic to do so.

The ideas described here are an attempt to develop an approach to argumentation that may be used in the interests of both parties. It aims to achieve this by unearthing the 'top layer' of the partner's reasoning apparatus and by attempting to work with it rather than against it. To this end, the utterances produced aim to influence the partner to believe what we believe to be in his best interests — although it may not be in fact. The utterances aim to convey what is so, and not to point out "where the partner is wrong". In the long term, this behaviour is intended to lead to the development of lasting relationships between agents that are underpinned both by the knowledge that their partners "treat them well" and that their partners act as they do "for the right reasons".

The ideas in this paper have been developed within a highly restrictive scenario that is deliberately asymmetric (being based on a buy / seller relationship). The structure of the analysis is far more general and applies to any scenario in which something has to be bought/made/designed that satisfies a need, and that can do various things. The agents who try to make it do things (use-cases if you like) subjectively rate what they see.

In previous work [3] we have advocated the gradual development of trust and intimacy[10] through successive argumentative exchanges as a way of building relationships between agents. The act of passing private information carries with it a sense of trust of the sender in the receiver, and having done so the sender will wish to observe that the receiver respects the information received. In this paper we have gone one step further by including a modest degree of *understanding* in the sense that an agent attempts to understand what her partner likes. This falls well short of a deep model of the partner's reasoning but we believe strikes a reasonable balance between being meaningful and being achievable. This augments the tools for building social relationships through argumentation by establishing:

- *trust* — my belief in the veracity of your *commitments*
- *intimacy* — my belief in the extent to which I know your private *information*
- *understanding* — my belief in the extent to which I know what you *like*

---

[10] The revelation of private information.

Acknowledgements. This research has been supported by the Sabbatical Programme of the Catalan Government BE2007, the Australian Research Council Discovery Grant DP0557168, and by the Spanish Ministerio de Educación y Ciencia project "Agreement Technologies" (CONSOLIDER CSD2007-0022, INGENIO 2010).

References

1. Sierra, C., Debenham, J.: Information-based agency. In: Proceedings of Twentieth International Joint Conference on Artificial Intelligence, IJCAI 2007, Hyderabad, India, pp. 1513–1518 (2007)
2. Sierra, C., Debenham, J.: Trust and honour in information-based agency. In: Stone, P., Weiss, G. (eds.) Proceedings Fifth International Conference on Autonomous Agents and Multi Agent Systems AAMAS 2006, Hakodate, Japan, pp. 1225–1232. ACM Press, New York (2006)
3. Sierra, C., Debenham, J.: The LOGIC Negotiation Model. In: Proceedings Sixth International Conference on Autonomous Agents and Multi Agent Systems AAMAS 2007, Honolulu, Hawai'i, pp. 1026–1033 (2007)
4. Dung, P.M.: On the Acceptability of Arguments and its Fundamental Role in Nonmonotonic Reasoning, Logic Programming and n-Person Games. Artificial Intelligence 77, 321–358 (1995)
5. Rahwan, I., Ramchurn, S., Jennings, N., McBurney, P., Parsons, S., Sonenberg, E.: Argumentation-based negotiation. Knowledge Engineering Review 18, 343–375 (2003)
6. Cheeseman, P., Stutz, J.: On The Relationship between Bayesian and Maximum Entropy Inference. In: Bayesian Inference and Maximum Entropy Methods in Science and Engineering, pp. 445–461. American Institute of Physics, Melville (2004)
7. Paris, J.: Common sense and maximum entropy. Synthese 117, 75–93 (1999)
8. Jaynes, E.: Information theory and statistical mechanics: Part I. Physical Review 106, 620–630 (1957)
9. Li, Y., Bandar, Z.A., McLean, D.: An approach for measuring semantic similarity between words using multiple information sources. IEEE Transactions on Knowledge and Data Engineering 15, 871–882 (2003)
10. Lewicki, R.J., Saunders, D.M., Minton, J.W.: Essentials of Negotiation. McGraw Hill, New York (2001)
11. Bentahar, J., Mbarki, M., Meyer, J.J.C., Moulin, B.: Strategic agent communication: An argumentation-driven approach. In: Baldoni, M., Son, T.C., van Riemsdijk, M.B., Winikoff, M. (eds.) DALT 2008. LNCS (LNAI), vol. 5397, pp. 233–250. Springer, Heidelberg (2009)
12. Kakas, A., Maudet, N., Moraitis, P.: Layered strategies and protocols layered strategies and protocols for argumentation-based agent interaction. In: Rahwan, I., Moraïtis, P., Reed, C. (eds.) ArgMAS 2004. LNCS (LNAI), vol. 3366, pp. 64–77. Springer, Heidelberg (2005)
13. Mbarki, M., Bentahar, J., Moulin, B.: Specification and complexity of strategic-based reasoning using argumentation. In: Maudet, N., Parsons, S., Rahwan, I. (eds.) ArgMAS 2006. LNCS (LNAI), vol. 4766, pp. 142–160. Springer, Heidelberg (2007)
14. Maslow, A.H.: A theory of human motivation. Psychological Review 50, 370–396 (1943)

Mediation = Information Revelation + Analogical Reasoning

Simeon Simoff[1,3], Carles Sierra[2,3], and Ramon López de Màntaras[2]

[1] School of Computing and Mathematics,
University of Western Sydney, NSW 1797, Australia
s.simoff@uws.edu.au
[2] Institut d'Investigació en Intel·ligència Artificial – IIIA,
Spanish Scientific Research Council, CSIC
08193 Bellaterra, Catalonia, Spain
{sierra, mantaras}@iiia.csic.es
[3] University of Technology, Sydney, Australia

Abstract. This paper presents an initial study of the relevant issues on the development of an automated mediation agent. The work is conducted within the 'curious negotiator' framework [1]. The paper demonstrates that mediation is a knowledge intensive process that integrates information revelation and analogical reasoning. The introduced formalism is used to demonstrate how via revealing the appropriate information and reshaping the set of issues of the disputing parties mediation can succeed. The paper presents *MediaThor* - a mediating agent that utilises past experiences and information from negotiating parties to mediate disputes and change the positions of negotiating parties.

1 Introduction

Negotiation is the process whereby two (or more) individual agents with conflicting interests interact, aiming at reaching a mutually beneficial agreement on a set of issues. Engaging in such interactions is a daily activity — from a simple negotiation on the price of a product we buy at the market to the complicated negotiations in dispute resolutions on the international arena. During such interactions, participants may need to make concessions in order to reach an agreement [2]. Individual negotiators usually have agendas of their own, which may be incompatible — there may be no solution that satisfies them all. Further the existence of a solution is unlikely to be known when the negotiation commences [3]. So it may not be useful to consider negotiation as a search problem because the solution space may be empty whilst the negotiating agents may believe that it is not so. If the negotiation is a multi-issue negotiation for which the issue set can change at any stage in the negotiation then the agendas of the individual negotiating agents must necessarily be at a higher level than the issues because the issues are unknown, and may even be issues that 'had never occurred' to one of the agents. Therefore in such cases the agendas of the agents cannot be a high level goal such as 'to maximise profit on the deal' as the deal space is unknown. Environmental conflict resolution is a typical example where conflicts involve many different types of parties, issues and resources and the issue set can change during the process [4].

J.-J.Ch. Meyer and J.M. Broersen (Eds.): KRAMAS 2008, LNAI 5605, pp. 145–160, 2009.

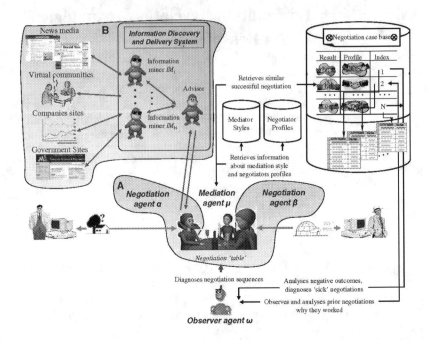

Fig. 1. The design of the curious negotiator and the progress of the research

As a result negotiations may reach a deadlock, taking prohibitively long time without reaching tangible outcomes, or be terminated. This is when in real life the intervention of a mediator can influence the process, facilitating it towards a mutual agreement.

The design of the 'curious negotiator' automated negotiation system, founded on the intuition "it's what you know that matters" [1], is an attempt to address these issues. Figure 1 shows an updated version of the overall design proposed in [1] and the progress of the work. Negotiation agent α negotiates with agent β by sending illocutions which represent offers and counter offers. The illocutions are represented in a communication language $\mathbb{C}$ that enable persuasive negotiation [5] and argumentation [6]. The research outcome under the shaded area A in Figure 1- the information-based agency [6], treats negotiation as an information discovery and revelation process. Such information may come from at least two sources: (i) from the 'negotiation table', e.g. from all utterances agents make during a negotiation [7]; and (ii) from various external electronic sources on the Internet. The research outcome under the shaded area B in Figure 1- the information discovery and delivery system that services negotiating agents, extracts information from various web sources [8] and provides recommendations to the negotiation agents in requested form (for example, refined exchange rate predictions [9] based on information from latest financial news; recommendations on product choice based on information extracted from consumer product reviews [10]). The mechanisms for dealing with negotiations that fail in reaching an agreement, or seemed to be leading to a failure, remain the undeveloped part of the 'curious negotiator' - the unshaded part in Figure 1. It includes the mediating agent μ, the observer agent ω and their supporting knowledge representation structures.

This paper presents the initial work on the principles of building an automated mediation agent within the 'curious negotiator' framework. We explore mediation as a process of intertwined information revelation and analogical reasoning, which incorporates past experiences. To model mediation the work draws from the research in dispute resolution. It specifies the requirements towards the knowledge representation structures supporting mediation. Section 2 looks at mediation, as a knowledge-driven process and explores the changes that information revelation can make to the negotiation space and the outcomes of negotiation. It introduces the notion of 'mental model' of participants involved in the process and looks at mechanisms of how these models can be utilised in automated mediation. Section 3 considers some aspects in utilising past experiences and background knowledge in automated mediation. It looks also at the utilisation of information at the diagnosis stage. Section 4 presents the architecture of MediaThor - a mediation agent that illustrates the computational implementation of mediation mechanisms discussed in the previous sections.

2   Mediation as a Knowledge Driven Process of Information Revelation

Contemporary analysts in social and political sciences look at mediation as a process that enables conflict resolution. Mediators are often indispensable in the area of *dispute (or conflict) resolutions*, settling variety of disputes, spanning from conflicts between sovereign nations to conflicts between family members, friends, and colleagues. Successful mediation can make a dramatic difference to the outcome of a negotiation stalemate. For instance, on 14 January 1998 the President of United Nations Security Council issued a statement demanding "that Iraq cooperate fully and immediately and without conditions with the Special Commission in accordance with the relevant resolutions."[1] As all UN weapons inspections in Iraq were frozen, during the following month all direct negotiations between the US and Iraq did not reach any agreement and the military conflict seemed unavoidable. The following event sequence illustrates the mediation process: (i) the US authorised the mediation effort; (ii) the UN secretary (the mediator) achieved a possible deal with Iraq; (iii) the UN secretary passed it back to the US; (iv) the US reviewed and accepted the deal. Several months later the conflict escalated, but this time no mediation was sought and military actions started. The mediation made a huge difference in the first dispute resolution.

2.1   Necessary and Sufficient Conditions for Mediation

This example illustrates that mediation as a process involves *information revelation* and part of the mediator's strategy is guiding the process of information revelation. The following are the necessary (C1, C2) and sufficient (C3) conditions for a mediation to take place:

- *Condition C1*: Negotiating agents α and β are willing to achieve a mutually beneficial agreement;

---

[1] http://daccessdds.un.org/doc/UNDOC/GEN/N98/007/84/PDF/N9800784.pdf

- *Condition C2*: Negotiating agents α and β are seeking or will accept mediation (in the first case, the awareness about the conflict and the problem with the current state of the negotiation resides with the negotiating agents, in the second case either the mediator agent μ or, if present, the observer agent ω diagnoses the problem);
- *Condition C3*: A mediating agent μ is available (this condition is by default embedded in the 'curious negotiator' paradigm).

In the example with the 1998 Iraq crisis, in the second case condition C2 was not present. Conflicts may be a result of a contradiction of interests, as in the example with the 1998 Iraqi crisis, but can be also a result just of a different (but unknown to the disputing parties) perception of the disputed subject.

2.2   Mediation Process within the 'Curious Negotiator' Framework

Further we consider the following mediation process, shown in Figure 2, where agents α and β are in a deadlock and direct exchange of offers between them has ceased. In a mediation session, α and β interact with messages m only with the mediating agent μ.

$\mathcal{M}^t$ denotes a "mental model" at time t. We use the label "mental model" to denote the view (including related knowledge) of an agent about a dispute, about the views of the other parties on that dispute and the expected outcomes. This knowledge is internal to the agent. Each model is manifested to the other agents through the actions taken by the agent. The label "mental models" has been chosen to emphasise the key role of the mediator in the careful examination of the way negotiation parties have built their views on the disputed issues [11]. It is also in accordance with the view that negotiation can be conceptualised as a problem-solving enterprise in which mental models guide the behaviour of negotiating parties [12]. Further in the text we use the term mental model without quotation marks.

$\mathcal{M}_\alpha^t$ and $\mathcal{M}_\beta^t$ denote the mental models of agents α and β, respectively. $\mathcal{M}_\alpha^t$ is not known by β and $\mathcal{M}_\beta^t$ is not known by α. None of them is known by the mediating agent μ. Each of these agents has its own approximations of the mental models of the other agents. $\mathcal{M}_{agent(party)}^t$ denotes the mental model that the *agent* has about another *party*. In particular, $\mathcal{M}_{\alpha(\beta)}^t$ is the mental model of α about β, i.e. about what β wants out of the negotiation; respectively, $\mathcal{M}_{\beta(\alpha)}^t$ is the mental model of β about α, i.e. the position of α in the dispute. Further, $\mathcal{M}_{\mu(\alpha)}^t$ and $\mathcal{M}_{\mu(\beta)}^t$ are the mental models of the mediating agent μ about the positions of α and β in the dispute, respectively.

We use the above formalism to demonstrate some aspects of mediation that need to be taken into account when developing automated mediators. Further we use two examples - The Orange Dispute [13] and the Sinai Peninsula Dispute to illustrate the role of information revelation and identification of analogy between disputes in order to reshape the set of issues and complete the mediation process.

2.3   The Orange Dispute - Reshaping the Problem Based on Additional Information

In the Orange Dispute [13], two sisters want the same orange. According to Kolodner [13] "MEDIATOR assumes they both want to eat it and solves the problem by having

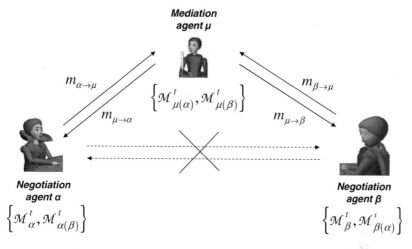

Fig. 2. The mediation agent within the 'curious negotiator' framework

one sister cut the orange in two and the second chooses her half. When the second sister uses her peel for baking and throws away the pulp, MEDIATOR realises it made a mistake."[2]

Further we present the two mediation attempts in terms of the agreements reached and the information that can be passed to the mediator. Lets our agent α represent the first sister who wants to have the orange as a desert and agent β represent the second sister who wants to have (only the peel of) the orange for cooking (the recipe requires the whole peel). If our mediation agent μ happens to be the case-based MEDIATOR, then the situation described in the Orange Dispute can be expressed through the mental models of the individual participants in Figure 3, making explicit the wrong assumption (the boxed expressions in Figure 3).

In these models t_{break}, and t_{start} indicate the time when negotiation broke and when mediation started, respectively (in the case of the MEDIATOR it has been a one step act). The results of the agreements in terms of the outcomes - *Outcome (agent, issue, result)* are presented in Table 1, where result values are denoted as follows: "+", "+/-" and "-" for positive, acceptable, and negative, respectively for the corresponding agents in terms of the negotiated issue. In the original example [13], the result in the outcome for β should be "+/-" as the second sister still used the peel from her half. Here we added the constraint of the recipe in order to get negotiation about the orange to a halt with an unacceptable "-" result and generate a request for mediation.

The Orange Dispute can be considered an example of a dispute over resource scarcity. The resource in this case has a possible component-based separation (without change of the total amount of available resource) that allows to change the structure of the dispute through mediation, opening the space for a mutually beneficial solution. It exposes two aspects of mediation:

---

[2] MEDIATOR [14] is one of the early case-based mediators. The focus of the work was on the use of case-based reasoning for problem understanding, solution generation, and failure recovery. The failure recovery ability is demonstrated with the Orange Dispute in [13].

$$\alpha \ wants \ the \ orange \ as \ a \ dessert \in \mathcal{M}_\alpha^t \tag{1}$$

$$\beta \ wants \ the \ peel \ of \ the \ orange \ for \ cooking \in \mathcal{M}_\beta^t \tag{2}$$

$$\beta \ wants \ an \ orange \in \mathcal{M}_{\alpha(\beta)}^{t_{break}} \tag{3}$$

$$\alpha \ wants \ an \ orange \in \mathcal{M}_{\beta(\alpha)}^{t_{break}} \tag{4}$$

$$\boxed{\alpha \ wants \ the \ orange \ as \ a \ dessert} \in \mathcal{M}_{\mu(\alpha)}^{t_{start}} \tag{5}$$

$$\boxed{\beta \ wants \ the \ orange \ as \ a \ dessert} \in \mathcal{M}_{\mu(\beta)}^{t_{start}} \tag{6}$$

Fig. 3. The wrong initial assumption of the MEDIATOR [14] in terms of our mental models (Boxed expressions). This initial assumption (which didn't change as there were no mechanisms for that) caused the failure of that mediator.

- The difference that a mediator can bring is in exploring the structure of the problem from a broader stance;
- An initial assumption by a mediator can lead to a failure of the mediation effort.

Consequently, we formulate the following postulates for the automated mediator:

- *Postulate P1*: An automated mediator μ should start interaction with extracting more information about the position of the parties on the negotiation;
- *Postulate P2*: An automated mediator should develop an independent "grand view" of the problem, which is more comprehensive than the individual views of α and β, respectively.;
- *Postulate P3*: An automated mediator μ should operate from the initial stance only of conditions C1 and C2.

Starting mediation without initial assumptions means that μ either does not have a model for each of the negotiating agents α and β, or accepts the models $\mathcal{M}_{\alpha(\beta)}^{t_{break}}$ and $\mathcal{M}_{\beta(\alpha)}^{t_{break}}$ these agents have about each other at the point of requesting mediation. In the case of the Orange Dispute, μ starts mediation with the exit models of α and β:

- $\mathcal{M}_{\mu(\alpha)}^{t_{start}} = \mathcal{M}_{\beta(\alpha)}^{t_{break}}$, i.e. α wants an orange $\in \mathcal{M}_{\mu(\alpha)}^{t_{start}}$, and
- $\mathcal{M}_{\mu(\beta)}^{t_{start}} = \mathcal{M}_{\alpha(\beta)}^{t_{break}}$, i.e. β wants an orange $\in \mathcal{M}_{\mu(\beta)}^{t_{start}}$.

This information is not sufficient for mediation, e.g. the uncertainty in the mutual models of α and β, and the model μ are the same. Research in conflict resolution in international relations demonstrates that if a mediator could credibly add information to the system of negotiators this alters the state of the system [15]. Consequently, μ takes

Table 1. Outcomes of the Orange Dispute, based on mediation with initial assumption

Agent	Agreement clauses	Outcome for α	Outcome for β
α	Cuts the orange into halves	Outcome(α, has orange, +/-)	Outcome(β, has orange, -)
β	chooses one half	Outcome(α, has orange, +/-)	Outcome(β, has orange, -)

$$\alpha \text{ wants the orange as a dessert} \in \mathcal{M}_\alpha^t \tag{7}$$

$$\beta \text{ wants the peel of the orange for cooking} \in \mathcal{M}_\beta^t \tag{8}$$

$$\beta \text{ wants an orange} \in \mathcal{M}_{\alpha(\beta)}^{t_{break}} \tag{9}$$

$$\alpha \text{ wants an orange} \in \mathcal{M}_{\beta(\alpha)}^{t_{break}} \tag{10}$$

$$\boxed{\alpha \text{ wants an orange}} \in \mathcal{M}_{\mu(\alpha)}^{t_{start}} \tag{11}$$

$$\boxed{\beta \text{ wants an orange}} \in \mathcal{M}_{\mu(\beta)}^{t_{start}} \tag{12}$$

$$\boxed{\alpha \text{ wants the orange as a dessert}} \in \mathcal{M}_{\mu(\alpha)}^{t_{end}} \tag{13}$$

$$\boxed{\beta \text{ wants the peel of the orange for cooking}} \in \mathcal{M}_{\mu(\beta)}^{t_{end}} \tag{14}$$

Fig. 4. The respective mental models of α, β and μ in the mediation session of the Orange Dispute with our proposed agent

steps in order to decrease this uncertainty. In addition, intuitively, it seems worth checking whether both parties have the same understanding of the issues in the dispute, i.e. technically, whether they operate with the same ontology or with compatible ontologies. In the Orange Dispute, μ obtains from each party what the orange is needed for. The Orange Dispute in terms of the mental models of the individual participants in the case of proposed mediation agent is presented in Figure 4. In these models t_{break}, t_{start} and t_{end} indicate the time when negotiation broke and when mediation started and ended, respectively. Note the difference of $\mathcal{M}_{\mu(\cdot)}^{t_{start}}$ for both α and β in Figure 3 and Figure 4. The steps taken by the mediating agent are described in Figure 5 (we do not use a formal illocution based language, but the actions that the language should cater for are shown in italic).

The Orange Dispute illustrates also another important ability that an automated mediator should posses — the ability to reshape or restructure the dispute, based on the additional information about the models of each party. The outcomes of the restructured Orange Dispute are shown in Table 2.

2.4   The Sinai Peninsula Dispute and Its Analogy with the Orange Dispute

The ability to reshape the problem is crucial for developing successful automated mediators. The Sinai Peninsula Dispute in the area of international relations shows similar properties to the Orange Dispute. The Sinai Peninsula is a piece of land of about 62,000 square km that separates Israel and Egypt. With its landscape Sinai has a *military value* for either side in terms of mechanised infantry transport or as a shelter for guerrilla

Table 2. Outcomes of the restructured Orange Dispute

Agent	Agreement clauses	Outcome for α	Outcome for β
α	Peels the orange	Outcome(α, eat, +)	Outcome(β, cook, +)
β	Gets the whole peel	Outcome(α, eat, +)	Outcome(β, cook, +)

1. μ : *ask* α to *send* its ontology of the negotiated item (orange).
2. μ : *ask* β to *send* its ontology of the negotiated item (orange).
3. μ : *compare* ontologies received from α and β.
4. μ : *send* α and β agreed ontology (orange as a fruit which has pulp and peel).
5. μ : *ask* α to *send* μ its preferences on the negotiated item in terms of agreed ontology.
6. μ : *ask* β to *send* μ its preferences on the negotiated item in terms of agreed ontology.
7. μ : *advises* α and β on $\mathcal{M}_\alpha^t$ and $\mathcal{M}_\beta^t$ based on their preferences
8. μ : *checks* the case base for past cases (resource disputes)
9. μ : *retrieves* resource disputes with divisible components
10. μ : *sends* α and β action separate resource (peel the orange)
11. μ : *tells* α and β to complete negotiation.
12. μ : mediation completed.

Fig. 5. Mediation as information revelation aiming at decreasing uncertainty within the negotiation system

forces. The perceived importance of the territory is evidenced by the fact that Israelis and Egyptians fought in or over the Sinai Peninsula in 1948, 1956, 1967, 1968-1970, and 1973. Since 1967 Sinai had been occupied by Israel. Figure 6 shows a very simplified version of the models of the parties at the initial meeting in Jerusalem, when the negotiations started and halted and the change of the mediators models that lead to the outcomes. For the purpose of this paper we aim to emphasise the high level analogy with the Orange Dispute case (see Figure 4), i.e. the need for a mediator to reframe the problem. In fact, the need for restructuring the problem in order for a mediator to get a "bigger picture" has been recognised in PERSUADER [2], to resolve labor-management disputes. In recent works [16] the mediator is expected to have a complete knowledge of the solution space.

Following the initial interaction in Jerusalem, the US President Jimmy Carter initiated a *third-party mediation* effort that culminated in the Camp David accords. For the purposes of this paper we consider a simplified version of the second agreement of the Camp David accords on the future of the Sinai Peninsula. The items in the agreement

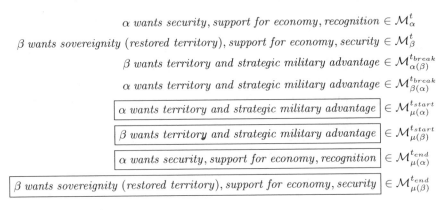

Fig. 6. The respective mental models of α, β and μ in the mediation session of the Sinai Dispute with our proposed agent

Table 3. The Sinai Peninsula Dispute. α denotes Israel; β denotes Egypt

Agent	Agreement clauses	Outcome for α	Outcome for β
α	withdraw its armed forces from the Sinai	Outcome(α, Military, -)	Outcome(β, Territory, +) Outcome(β, Sovereignty, +)
α	Evacuate its 4500 civilians	Outcome(α, Territory, -)	Outcome(β, Territory, +) Outcome(β, Sovereignty, +)
α	Restore Sinai to Egypt	Outcome(α, Territory, -)	Outcome(β, Territory, +) Outcome(β, Sovereignty, +)
α	Limit its forces within 3km from Egyptian Border	Outcome(α, Military, -) Outcome(α, Security, +)	Outcome(β, Security, +)
α	Lost the Abu-Rudeis oil fields in Western Sinai	Outcome(α, Economy, -)	Outcome(β, Economy, +)
β	Normal diplomatic relations with Israel	Outcome(α, Recognition, +)	Outcome(β, Security, +)
β	Freedom of passage through Suez Canal	Outcome(α, Economy, +) Outcome(α, Security, +)	Outcome(β, Security, +)
β	Freedom of passage through nearby waters	Outcome(α, Economy, +) Outcome(α, Security, +)	Outcome(β, Economy, +) Outcome(β, Security, +)
β	Restricted Egyptian forces in Sinai	Outcome(α, Security, +)	Outcome(β, Military, -) Outcome(β, Security, +)

are presented in Table 3, in a structure, similar to the presentation of the agreements in the Orange Dispute in Table 1 and Table 2. Without getting into the details of the mediation steps, from Table 3 it is evidenced that the initial mutually perceived models and about the need for territory and strategic military advantage have been transformed by the mediation into a Security/Sovereignty trade-off, with economic benefits.

The analogy with the Orange Dispute is in having the initial negotiation framed around a common resource Territory and a similar issue of having strategic military advantage as the main goals that can enable the security. Though both territorial and military components remain on the negotiation table, based on some background knowledge and higher level view of the ontology of the problem, the mediator developed a view of the set of issues and aligned the ontologies of both parties which eventually changed their models: security and restoration may not necessarily be achieved with occupation of a territory or with expensive military presence.

The information injected by the mediator and proposed steps leads to decreasing the differences between perceived mental models $\mathcal{M}^t_{\alpha(\beta)}$ and $\mathcal{M}^t_{\beta(\alpha)}$, and the corresponding actual mental models $\mathcal{M}^t_\beta$ and $\mathcal{M}^t_\alpha$ of agents α and β, respectively, i.e. the intervention of the mediator decreases the uncertainty in the negotiation system.

3   Utilising Past Experiences and Background Knowledge in Automated Mediation

The American Bar Association defines mediation as a process by which those who have a dispute, misunderstanding or conflict come together and, with *the assistance of a trained neutral mediator*, resolve the issues and problems in a way that meets the needs and interests of both parties.[3] This definition emphasises the unbiased nature of the mediator and the key role of its past experience. The *bias of a mediator* is defined as the presence of a preference towards one of the outcomes in the negotiation; or, sides involved in the negotiation. Not having preference towards any of the outcomes of a negotiation means also to keep open all options. For instance, the peace-loving broker's bias towards peaceful solutions makes his or her claims less believable compared to a broker who is indifferent to war or peace [15]. Such bias as a result can decrease the effectiveness of the mediation effort. Protecting automated mediation from introduction of a bias is not seen as a problem.

Experience is, perhaps, the distinct feature between successful and less successful mediators. Analogical reasoning (CBR + ontology) is an approach to problem solving that emphasizes the role of prior experience during future problem solving (i.e., new problems are solved by reusing and if necessary adapting the solutions to similar problems that were solved in the past) (see [17] for a recent review of the state-of-the-art in the CBR field). From a machine learning point of view, updating the case base learning without generalisation. Some aspects of using the past experience by the tandem Mediation and Observation agents have been discussed in [1]. In terms of required case representation, a starting point is the knowledge representation structure for representing negotiation cases, proposed in [18]. This structure needs to be updated for dealing with ontologies. For the mediation, the case based will be linked to the corresponding knowledge base of the mediation strategies used. The case structure now includes a negotiation case as its problem section and the collection of mediation steps, information used and other knowledge, as the solution part of the case.

Important from a computational perspective is the diagnosis stage of the mediation process [19]. The diagnostic function consists of monitoring the progress of negotiation or related interactions intended to settle or resolve disputed issues (Druckman and co-authors [19] refer to [20]). Monitoring provides a long view of unfolding developments, including trends in escalating and de-escalating dynamics. Within the framework of 'curious negotiator' we consider this stage as a pre-mediation stage, which is executed by the observer agent ω. To some extent it resembles similarity with OLAP[4] — the pre-data mining steps in business intelligence, where summary statistics at different levels are generated and later provide guidance to the data mining strategies. Similar to OLAP, monitoring should be able to provide snapshots of the negotiation process at any moment of time at different levels of granularity. The mediator μ should be able to estimate the difference between $\mathcal{M}^t_{\alpha(\beta)}$ and $\mathcal{M}^t_{\beta(\alpha)}$ from the respective actual mental models $\mathcal{M}^t_\beta$ and $\mathcal{M}^t_\alpha$ in order to define the intervention time of mediating interventions (if we follow a proactive approach and intervene before negotiation fails).

4   MediaThor: A Powerful CBR Mediator Agent

The architecture of a mediator agent as described in the previous sections has to be based on a clear understanding of the relationships between its models of the agents

---

[3] http://www.abanet.org/cpr/clientpro/medpreface.html
[4] Online analytical processing.

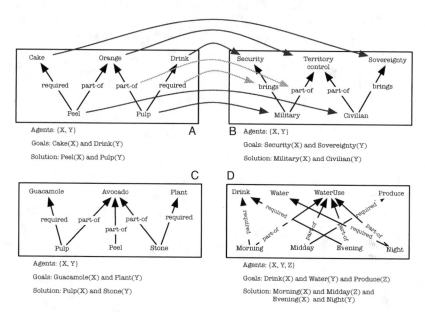

Fig. 7. Four cases represented as: an ontology (a graph), the agents participating and the goals and solution of the problem. Blue arrows represent a structural matching between the ontologies of cases A and B, while green arrows represent a semantic matching between the same cases.

and the ontology that describes the problem and its solution. The retrieval of a previous case, described in completely different ontological terms, has to rely on some alignment process between the concepts and semantic relations of that past case and the concepts and relations in the problem being solved. Figure 7 shows a representation of four cases. Each case consists of an ontology representing the problem, the agents participating, the goals each agent had, and the solution found. On Figure 7-A and 7-B one can see a representation of the orange and Sinai peninsula disputes. We consider two types of ontology matching in this architecture. A structural matching that aims at connecting the nodes in the ontology from a structural point of view (arrows in blue in the picture — or dark grey for the B&W printers) and from a semantic point of view (arrows in green —or in light grey). The example would account for a perfect matching from a structural point of view and for an almost perfect matching from a semantic point of view (up to the semantic similarity between *required* and *brings*.)

Thus, assuming that the set of all possible ontologies[5] is noted by $\mathcal{O}$, we will therefore require that any mediator is provided with a $Match : \mathcal{O} \times \mathcal{O} \rightarrow [0, 1]$ function that determines the level of similarity between any two ontologies (graphs). This function will be a combination of at least structural and semantic similarities between the ontologies. A number of existing solutions can be found in [21]. We will also assume that the ontology of a case is indeed a partial view (subgraph) of a more general ontology. Thus, given an ontology $o \in \mathcal{O}$ we will note by $\bar{o}$ this general ontology. In our example, the

---

[5] In this paper we will take the simplified view that ontologies are labelled directed graphs, with concepts at the nodes and binary relations on the links.

orange case could be seen as a view of an ontology of fruits and their usage, while the Sinai case could be seen as a view of an ontology on military affairs. We will therefore associate to each case the ontology from where the case is drawn and moreover we will define a subsumption relation between ontologies, $\sqsubseteq \subseteq \mathcal{O} \times \mathcal{O}$, being $o \sqsubseteq o'$ true when o is a sub-ontology of o', i.e. a subgraph. Clearly $o_i \sqsubseteq \bar{o}_i$

The concepts in the ontologies will all be understood as resources, and we abuse notation by transforming them into predicates with the following intuitive meaning: $Security(X)$ means that agent X *gets Security*. The difference between goals and solutions in the cases have to be understood from a planning perspective: a solution is an assignment of resources to agents that logically imply their respective goals.

According to our view above, *MediaThor* will take a very specific view on what $\mathcal{M}^t_{\mu(\cdot)}$ is. In particular we will assume that $\mathcal{M}^t_{\mu(\alpha)} = \langle o_\alpha, g_\alpha, r_\alpha \rangle$ where o_α, g_α, and r_α is the perceived ontology, goals, and reservations of α. We understand by reservations those constraints that α requires the solution to satisfy. *MediaThor* is equipped with a predicate $Sat \subseteq \mathcal{C} \times 2^{\mathcal{R}}$ that is true when a particular case satisfies a set of reservations.

We describe the architecture of *MediaThor* following the steps of a CBR algorithm.

Retrieval. *MediaThor* has a memory of past cases $C = \{c_i\}_{i \in I}$ where $c_i = \langle A_i, o_i, G_i, S_i \rangle$, A_i the set of agents involved, o_i an ontology, G_i the goals of the agents participating in the case, and S_i the solution as a set or resource assignments. *MediaThor* also is dealing with agents α and β trying to find a solution to their problem. The model that *MediaThor* has at any moment of time is a partial case where the ontology may be incomplete, and the goals may be wrong. Thus, the current problem is a case c using ontology o.

MediaThor, at retrieval time, looks for cases that are similar to the current case c. *MediaThor* uses a sliding parameter η that filters those cases c_i that are expressed in ontologies o_i that are semantically similar to o up to the value η. By setting $\eta = 1$ we are considering cases exactly over the same ontology, and by decreasing η we consider increasingly farther away cases in terms of semantic similarity.

Figure 7 shows that there is a structural mapping between cases B and C, although in this case C contains a richer structure. Ontology matchers provide a degree of matching but also a function that maps the concepts and labels of one ontology into the concepts and labels of the other. Thus, if the mapping function from B to C is called $f_{B \to C}$ it is clear that $f_{B \to C}(o_B) \sqsubseteq o_C$ and that the degree of structural matching between A and C has to be smaller than between A and B. Thus assuming the semantic matching (S-Match) is the same between A and B and between C and B, we can conclude that if B is the current case *MediaThor* would prefer case A to case C at retrieval time.

Thus, for a given η, $\mathcal{M}^t_{\mu(\cdot)}$, current case $c = \langle \{\alpha, \beta\}, o, G_\alpha \cup G_\beta, \emptyset \rangle$ and case base C, the best case $c^* \in C$ is defined as:

$$c^* = arg \max_{\substack{c_i \in C \\ S-Match(\bar{o}_i, \bar{o}) \leq \eta \\ Sat(c_i, R_\alpha \cup R_\beta)}} Match(f_{c_i \to c}(o_i), \hat{o}_i)$$

where

$$\hat{o}_i = arg \max_{\substack{o' \\ o \sqsubseteq o' \sqsubseteq \bar{o}}} Match(f_{c_i \rightarrow c}(o_i), o')$$

Reuse. Adaptation is solved by local exploration around the case found while satisfying the reservations of the agents in conflict. The exploration is made by looking for expansions or contractions of the ontology of the case that might increase the matching degree with the current case, and then using the so expanded ontology to generate a solution.

$$o^R = arg \max_{\substack{o' \\ o^* \sqsubseteq o' \sqsubseteq \bar{o}^*}} Match(f_{c_i \rightarrow c}(o'), o)$$

Revise. Here the solution generated in the step above is proposed to the participants. In this way changes in the goal set of the agents or new reservations may appear. Also, the ontology can be refined as the adaptation of the ontology may have introduced elements found irrelevant by the participants. For instance in the case of the water dispute (Figure 7-D) a farmer may say that watering the fields on midday is unfeasible because of sun heat. This reservation might make a solution in which watering is done in the middle of the day unfeasible. If no succeess is obtained, another iteration is made.

$$R'_\alpha = f(Dialogue, R_\alpha)$$

Retain. *MediaThor* follows a simple method to memorise new cases. A case will be memorised if there is no other case in the memory of cases that has a similarity degree (using the matching between the ontologies) with it over a threshold θ. That is, if the memory of cases at time t is C^t and the solved case at time t is c then

$$C^{t+1} = \begin{cases} C^t \cup \{c\} & \text{if } Sim(c, c^*) < \theta \\ C^t & \text{otherwise} \end{cases} \tag{15}$$

We will update the ontologies as a fusion/combination of the ontologies of the cases. In that way we can avoid asking for an ontology which is always annoying. The details of the fusion/combination of ontologies are beyond the scope of this paper.

5   Conclusions

Though during the years there has been some interest in automated mediation [14,2,22,16], the field requires a significant effort in research and development. This paper has presented the initial work on the principles of building an automated mediation agent within the 'curious negotiator' framework, hence can utilise some of the machinery developed for it, in particular: (i) the *information-based agency* [7,6]; (ii) the *information-mining system* [8,9]; and (iii) the *electronic/virtual institutions environment* [23,24], which offers means for performing negotiation and collecting the necessary data about the negotiation sessions in order to use it in mediation.

We established that mediation is an information revelation process where analogies (including ones across different domains) can play key role in reshaping the set of issues. The Orange and Sinai disputes demonstrate how through the revealing of the appropriate information and applying analogy to reshape a dispute mediation can succeed. *MediaThor* demonstrated that computationally, the approach requires the specification of the introduced mental models of the agents and the mechanisms for aligning/agreeing on the ontologies of the dispute that they use. *MediaThor* also demonstrated that case-based reasoning offers a potential mechanism for the mediator for handling past experiences, though the structure of the case is complex (in comparison to the attribute-value structure), extending the already complex structure for representing negotiation cases [18]. Automating mediation must take in account that mediation is (i) a knowledge intensive process, where the mediators utilise their past experiences; and (ii) a process that utilises information from negotiating parties and uses information for changing the positions these parties have on the negotiation table.

As the mediator utilises information to decrease the uncertainty in the dispute, an automated mediation would require a measure of uncertainty $\mathbb{H}(\mathcal{M}^t)$, allowing to quantify and compare the uncertainty coming from the incomplete knowledge of the mental models of the agents. In terms of the two party mediation in Figure 2, this decrease should be observable, i.e. $\mathbb{H}(\mathcal{M}^t_{\mu(\alpha)}) < \mathbb{H}(\mathcal{M}^t_{\beta(\alpha)})$ and $\mathbb{H}(\mathcal{M}^t_{\mu(\beta)}) < \mathbb{H}(\mathcal{M}^t_{\alpha(\beta)})$. Within the framework of the information-theoretic approach, such measure should measure the "information gain", as the mediator adds such gain. Viewing mediation as a dialogue system (see Figure 2) points also to the information-theoretic work in dialogue management strategies in conversational case-based reasoning [25]. In terms of an automated mediation system, the mediator should have the mechanism to determine the most informative question to ask at each stage of the interaction to each of the negotiating agents. These issues remain beyond the scope of this paper.

Though beyond the scope of the paper, we are aware that mediation requires trust in the mediator from the parties involved, as much of the information about their position negotiating parties would not reveal to the other side.

In conclusion, we would like to note that nowadays mediation skills are taught to students at various levels and schools spanning from elementary schools to university schools, including the Harvard Law School. Hence, the development of an automated mediation system is on the top priority of the research agendas.

Acknowledgements

This research has been supported by the Sabbatical Progrmme of the Spanish Ministerio de Educación y Ciencia Grant SAB2006-0001, the Australian Research Council Discovery Grant DP0557168, the Generalitat de Catalunya Grant 2005/SGR/00093 and by the Spanish Ministerio de Educacion y Ciencia Consolider Grant CSD2007-0022.

References

1. Simoff, S.J., Debenham, J.: Curious negotiator. In: Klusch, M., Ossowski, S., Shehory, O. (eds.) CIA 2002. LNCS (LNAI), vol. 2446, p. 104. Springer, Heidelberg (2002)
2. Sycara, K.P.: Problem restructuring in negotiation. Management Science 37(10), 1248–1268 (1991)

3. Lewicki, R.J., Saunders, D.M., Minton, J.W.: Essentials of Negotiation. McGraw-Hill, New York (2001)
4. Franklin Dukes, E.: What we know about environmental conflict resolution: An analysis based on research. Conflict Resolution Quarterly 22(1-2), 191–220 (2004)
5. Ramchurn, S.D., Sierra, C., Godo, L., Jennings, N.R.: Negotiating using rewards. Artificial Intelligence 171, 805–837 (2007)
6. Sierra, C., Debenham, J.: Information-based agency. In: Proceedings of Twentieth International Joint Conference on Artificial Intelligence, IJCAI 2007, Hyderabad, India, pp. 1513–1518 (2007)
7. Debenham, J.: Bargaining with information. In: Jennings, N.R., Sierra, C., Sonenberg, L., Tambe, M. (eds.) Proceedings Third International Conference on Autonomous Agents and Multi Agent Systems AAMAS 2004, pp. 664–671. ACM Press, New York (2004)
8. Zhang, D., Simoff, S.: Informing the curious negotiator: Automatic news extraction from the internet. In: Simoff, S., Williams, G. (eds.) Proceedings 3rd Australasian Data Mining Conference, Cairns, Australia, December 6-7, pp. 55–72 (2004)
9. Zhang, D., Simoff, S., Debenham, J.: Exchange rate modelling for e-negotiators using text mining techniques. In: E-Service Intelligence - Methodologies, Technologies and Applications, pp. 191–211. Springer, Heidelberg (2007)
10. Aciar, S., Zhang, D., Simoff, S., Debenham, J.: Informed recommender: Basing recommendations on consumer product reviews. IEEE Intelligent Systems, 39–47 (May/June 2007)
11. Gentner, D., Stevens, A.L. (eds.): Mental Models. Erlbaum, Hillsdale (1983)
12. Van Boven, L., Thompson, L.: A look into the mind of the negotiator: Mental models in negotiation. Group Processes & Intergroup Relations 6(4), 387–404 (2003)
13. Kolodner, J.: Case-Based Reasoning. Morgan Kaufmann Publishers, Inc., San Mateo (1993)
14. Kolodner, J.L., Simpson, R.L.: The mediator: Analysis of an early case-based problem solver. Cognitive Science 13(4), 507–549 (1989)
15. Smith, A., Stam, A.: Mediation and peacekeeping in a random walk model of civil and interstate war. International Studies Review 5(4), 115–135 (2003)
16. Chalamish, M., Kraus, S.: Automed - an automated mediator for bilateral negotiations under time constraints. In: Proceedings of the International Conference on Autonomous Agents and Multi Agent Systems, AAMAS 2007, Hawaii, USA. IFAAMAS (2007)
17. De Mantaras, R.L., McSherry, D., Bridge, D., Leake, D., Smyth, B., Craw, S., Faltings, B., Maher, M.L., Cox, M.T., Forbus, K., Keane, M., Aamodt, A., Watson, I.: Retrieval, reuse, revision and retention in case-based reasoning. The Knowledge Engineering Review 20(3), 215–240 (2005)
18. Matos, N., Sierra, C.: Evolutionary computing and negotiating agents. In: Noriega, P., Sierra, C. (eds.) AMEC 1998. LNCS (LNAI), vol. 1571, pp. 126–150. Springer, Heidelberg (1999)
19. Druckman, D., Druckman, J.N., Arai, T.: e-mediation: Evaluating the impacts of an electronic mediator on negotiating behavior. Group Decision and Negotiation 13, 481–511 (2004)
20. Zartman, I.W., Berman, M.R.: The Practical Negotiator. Yale University Press, New Haven (1982)
21. Giunchiglia, F., Yatskevich, M., Shvaiko, P.: Semantic matching: Algorithms and implementation. In: Spaccapietra, S., Atzeni, P., Fages, F., Hacid, M.-S., Kifer, M., Mylopoulos, J., Pernici, B., Shvaiko, P., Trujillo, J., Zaihrayeu, I. (eds.) Journal on Data Semantics IX. LNCS, vol. 4601, pp. 1–38. Springer, Heidelberg (2007)
22. Wilkenfeld, J., Kraus, S., Santmire, T.E., Frain, C.K.: The role of mediation in conflict management: Conditions for successful resolution. In: Multiple Paths to Knowledge in International Relations. Lexington Books (2004)

23. Esteva, M.: Electronic Institutions: From specification to development. Phd thesis, Technical University of Catalonia, Barcelona (2003)
24. Bogdanovych, A.: Virtual Institutions. Phd thesis, Faculty of Information Technology, University of Technology, Sydney, Sydney (2007)
25. Branting, K., Lester, J., Mott, B.: Dialogue management for conversational case-based reasoning. In: Funk, P., González Calero, P.A. (eds.) ECCBR 2004. LNCS (LNAI), vol. 3155, pp. 77–90. Springer, Heidelberg (2004)

Author Index